Contents

1. Mathematical Tools Needed For The Understanding of Some Basic Concepts In This Book …… 2
2. Random Variable …… 20
3. Mean and Variance of Some Discrete Probability Distributions …… 43
4. Practical Application of Some Probability Distributions …… 61
5. Moment of a Random Variable (Discrete and Continuous) …… 95
6. Continuous Random Variable …… 111
7. Moment Generating Function (m.g.f) And Cummulant Generating Function …… 158
8. Bivariate Normal Distribution and Stochastic Independence …… 191
9. Central Limit Theorem and Application …… 206
10. Point Estimation …… 221
11. Theory of Hypothesis Testing …… 248

Chapter 1

Mathematical Tools Needed For The Understanding of Some Basic Concepts In This Book

Differentiation, Integration, and Series are the mathematical tools needed for the understanding of some basic concepts in this book.

Differentiation

Direct Differential Laws

(i) $y = x^n$, $\dfrac{dy}{dx} = nx^{n-1}$

(ii) $y = ax^n$, $\dfrac{dy}{dx} = anx^{n-1}$

(iii) $y = k$, $\dfrac{dy}{dx} = 0$ (k is an arbitrary constant)

Example

Differentiate the following functions

(i) $2x^3 + 3x^2 + 2x - 5$

Chapter 1

(ii) $7x - \dfrac{2}{x^2} + \dfrac{3}{x}$

(iii) $\dfrac{5x^5 + 4x^3 - 3x^2}{6x^2}$

(iv) $(3x^2 + 2x + 1)^2$

Solution

$$\begin{aligned}
\text{(i) Let } y &= 2x^3 + 3x^2 + 2x - 5 \\
\therefore \dfrac{dy}{dx} &= 6x^2 + 6x + 2
\end{aligned}$$

$$\begin{aligned}
\text{(ii) Let } y &= 7x - \dfrac{2}{x^2} + \dfrac{3}{x} \\
&= 7x - 2x^{-2} + 3x^{-1} \\
\therefore \dfrac{dy}{dx} &= 7 + 4x^{-3} - 3x^{-2} \\
&= 7 + \dfrac{4}{x^3} - \dfrac{3}{x^2}
\end{aligned}$$

$$\begin{aligned}
\text{(iii) Let } y &= \dfrac{5x^5 + 4x^3 - 3x^2}{6x^2} \\
&= \dfrac{5x^3}{6} + \dfrac{4x}{6} - \dfrac{3}{6} \\
\therefore \dfrac{dy}{dx} &= \dfrac{15x^2}{6} + \dfrac{4}{6} = \dfrac{5x^2}{2} + \dfrac{2}{3}
\end{aligned}$$

$$\begin{aligned}
\text{(iv) Let } y &= (3x^2 + 2x + 1)^2 \\
&= (3x^2 + 2x + 1)(3x^2 + 2x + 1) \\
&= 9x^4 + 12x^3 + 10x^2 + 4x + 1 \\
\therefore \dfrac{dy}{dx} &= 36x^3 + 36x^2 + 20x + 4 \\
&= 4(9x^3 + 9x^2 + 5x + 1)
\end{aligned}$$

Note:

(i) If the given expression has a common denominator, we divide each term by the common denominator before differentiating (see example (iii) above).

(ii) Apart from y, the function to be differentiated may be represented by any other letter (small or capital) except zero which can be confused with letter 0.

Exercise:
Differentiate the following:

(i) $2x^7 + 3x^5 - 2x^3 + 5x - 3$

(ii) $5x^2 + 4x - \dfrac{3}{x^2} - 7$

(iii) $\dfrac{4x^4 + 9x^3 + 3x^2 - 3}{3x}$

(iv) $(2x^2 + x + 3)^2$

Function of Functions

The task of expanding function like $(3x^2 + 2x + 1)^{70}$ becomes cumbersome and may not be error free except by calculating machine. Hence to differentiate this type of function, method of differentiation of Function of Function has to be applied.

Example:
Differentiate the following functions:

(i) $(cx^3 + bx + d)^n$

(ii) $(2x^2 + 3x + 1)^{70}$

(iii) $(x^2 - \dfrac{1}{x})^8$

Chapter 1 5

Solution:
Let $y = (cx^3 + bx + d)^n$

(i) $t = cx^3 + bx + d \qquad \dfrac{dt}{dx} = 3cx^2 + b$

$\therefore \ y = t^n \quad \therefore \ \dfrac{dy}{dt} = nt^{n-1}$

$\begin{aligned}\dfrac{dy}{dx} = \dfrac{dy}{dt} \cdot \dfrac{dt}{dx} &= nt^{n-1}(3cx^2 + b) \\ &= n(cx^3 + bx + d)^{n-1}(3cx^2 + b) \\ &= n(3cx^2 + b)(cx^3 + bx + d)^{n-1}\end{aligned}$

(ii) Let $y = (2x^2 + 3x + 1)^{70}$

$p = (2x^2 + 3x + 1) \quad \therefore \ \dfrac{dp}{dx} = 4x + 3$

$\therefore \ y = p^{70} \quad \dfrac{dy}{dp} = 70p^{69}$

$\begin{aligned}\dfrac{dy}{dx} = \dfrac{dy}{dp} \cdot \dfrac{dp}{dx} &= 70p^{69}(4x + 3) \\ &= 70(4x + 3)p^{69} \\ &= 70(4x + 3)(2x^2 + 3x + 1)^{69}\end{aligned}$

(iii) Let $y = \left(x^2 - \dfrac{1}{x}\right)^8$ and

$r = x^2 - \dfrac{1}{x} \Rightarrow r = x^2 - x^{-1}, \ \dfrac{dr}{dx} = 2x + x^{-2}$

$\therefore \ y = r^8, \ \dfrac{dy}{dr} = 8r^7,$

$\begin{aligned}\dfrac{dy}{dx} = \dfrac{dy}{dr} \times \dfrac{dr}{dx} &= 8r^7(2x + x^{-2}) \\ &= 8(2x + x^{-2})(x^2 - \dfrac{1}{x})^7 \\ &= 8(2x + \dfrac{1}{x^2})(x^2 - \dfrac{1}{x})^7\end{aligned}$

Chapter 1

Exercise
Differentiate the following:

(i) $(ax^3 + bx^2 + cx + k)^n$

(ii) $(8x^2 + 3x + 4)^{90}$

(iii) $\sqrt{2x^3 - \dfrac{1}{x} + 2x}$

(iv) $\left(x - \dfrac{1}{x}\right)^7$.

Alternative method

(i) Given that $y = (cx^3 + dx^2 + k)^n$, find $\dfrac{dy}{dx}$

$$\begin{aligned}\dfrac{dy}{dx} &= n(cx^3 + dx^2 + k)^{n-1} \dfrac{d}{dx}(cx^3 + dx^2 + k) \\ &= n(cx^3 + dx^2 + k)^{n-1}(3cx^2 + 2dx) \\ &= n(3cx^2 + 2dx)(cx^3 + dx^2 + k)^{n-1}\end{aligned}$$

(ii) Given that

$$\begin{aligned}y &= (6x^2 + 3x + 4)^{90} \\ \dfrac{dy}{dx} &= 90(6x^2 + 3x + 4)^{89} \dfrac{d}{dx}(6x^2 + 3x + 4) \\ &= 90(6x^2 + 3x + 4)^{89}(12x + 3) \\ &= 270(4x + 1)(6x^2 + 3x + 4)^{89}\end{aligned}$$

Exercises
Use the above method to differentiate the following:

(i) $(2x^2 + 3x + 2)^9$

(ii) $\left(x - \dfrac{1}{x}\right)^{\frac{1}{3}}$

Product Rule

Given that u and v are different functions of x and y where $y = uv$ is the product of u and v; by product rule, $\dfrac{dy}{dx} = u\dfrac{dv}{dx} + v\dfrac{du}{dx}$.

Example:
Find the derivative of the following functions:

(i) $(2x+3)(3x-4)$ (ii) $(2x+6)^2(5x-4)$ (iii) $(3x-5)\left(6 - \dfrac{2}{x}\right)$

Solution

(i) Put $y = (2x+3)(3x-4)$

Let $u = 2x + 3$ $\therefore \dfrac{du}{dx} = 2$

$v = 3x - 4$ $\therefore \dfrac{dv}{dx} = 3$

$$\begin{aligned}\dfrac{dy}{dx} = u\dfrac{dv}{dx} + v\dfrac{du}{dx} &= (2x+3)(3) + (3x-4)2 \\ &= 6x + 9 + 6x - 8 \\ &= 12x + 1\end{aligned}$$

(ii) Put $y = (2x+6)^2(5x-4)$

$u = (2x+6)^2$ $\therefore \dfrac{du}{dx} = 2(2x+6)2 = 4(2x+6)$

$v = 5x - 4$ $\dfrac{dv}{dx} = 5$

$$\begin{aligned}\therefore \dfrac{dy}{dx} = u\dfrac{dv}{dx} + v\dfrac{du}{dx} &= (2x+6)^2(5) + 4(5x-4)(2x+6) \\ &= (4x^2 + 24x + 36)5 + 4(10x^2 + 30x - 8x - 24) \\ &= 20x^2 + 120x + 180 + 40x^2 + 88x - 96. \\ &= 60x^2 + 208x + 84.\end{aligned}$$

(iii) Put $y = (3x-5)(6 - \frac{2}{x})$

$u = 3x - 5 \quad \therefore \frac{du}{dx} = 3$

$v = 6 - \frac{2}{x} = 6 - 2x^{-1} \quad \frac{dv}{dx} = 2x^{-2}$

$$\therefore \frac{dy}{dx} = (3x-5)(2x^{-2}) + (6 - 2x^{-1})3$$
$$= 6x^{-1} - 10x^{-2} + 18 - 6x^{-1}$$
$$= 18 - 10x^{-2} = 18 - \frac{10}{x^2} = \frac{18x^2 - 10}{x^2}$$

Exercises

Differentiate the following functions:

(i) $(2x^2 + x + 7)(x + 3)$ (ii) $(3x - 8)^2(2x + 5)$ (iii) $\left(x - \frac{2}{x}\right)(3x^2 + 5)$

Quotient Rule

Given that u and v are different functions of x such that $y = \frac{u}{v}$, by Quotient Rule, $\frac{dy}{dx} = \frac{v\frac{du}{dx} - u\frac{dv}{dx}}{v^2}$.

Example:

Differentiate the following:

(i) $\frac{7x - 8}{8x - 7}$ (ii) $\frac{(3x + 2)^2}{2x - 3}$ (iii) $\frac{4x^2 - 3}{2x + 5}$

Solution:

(i) Put $y = \frac{7x - 8}{8x - 7}$

$u = 7x - 8 \quad \therefore \frac{du}{dx} = 7$

$$v = 8x - 7 \quad \therefore \frac{dv}{dx} = 8$$

$$\therefore \frac{dy}{dx} = \frac{v\frac{du}{dx} - u\frac{dv}{dx}}{v^2} = \frac{(8x-7)(7) - (7x-8)8}{(8x-7)^2} = \frac{56x - 49 - 56x + 64}{(8x-7)^2} = \frac{15}{(8x-7)^2}$$

(ii) Put $y = \dfrac{(3x+2)^2}{2x-3}$

$u = (3x+2)^2 \qquad \dfrac{du}{dx} = 2(3)(3x+2) = 18x + 12$

$v = 2x - 3 \qquad \dfrac{dv}{dx} = 2$

$$\begin{aligned}\frac{dy}{dx} &= \frac{(2x-3)(18x+12) - (3x+2)^2(2)}{(2x-3)^2} \\ &= \frac{36x^2 + 24x - 54x - 36 - 2(9x^2 + 12x + 4)}{(2x-3)^2} \\ &= \frac{18x^2 - 54x - 44}{(2x-3)^2} = \frac{2(9x^2 - 27x - 22)}{(2x-3)^2}\end{aligned}$$

(iii) Put $y = \dfrac{4x^2 - 3}{3x + 5}$

$u = 4x^2 - 3, \qquad \therefore \dfrac{du}{dx} = 8x$

$v = 3x + 5 \qquad \dfrac{dv}{dx} = 3$

$$\begin{aligned}\therefore \frac{dy}{dx} &= \frac{(3x+5)(8x) - (4x^2 - 3)3}{(3x-15)^2} \\ &= \frac{24x^2 + 40x - 12x^2 + 9}{(3x+5)^2} = \frac{12x^2 + 40x + 9}{(3x+5)^2}\end{aligned}$$

Partial Derivative

Given a function $f(x, y, z)$, the partial derivative of $f(x, y, z)$ with respect to x, y, z respectively is given by $\dfrac{\partial f}{\partial x}, \dfrac{\partial f}{\partial y}, \dfrac{\partial f}{\partial z}$.

Example:
Given that $f(x,y,z) = 3x^3y - 2xy^2z$.
Find

(i) $\dfrac{\partial f}{\partial x}, \dfrac{\partial f}{\partial y}, \dfrac{\partial f}{\partial z}$ (ii) $\dfrac{\partial f}{\partial y} - \dfrac{\partial f}{\partial x}$

(iii) $\dfrac{\partial f}{\partial x} + \dfrac{\partial f}{\partial y} - \dfrac{\partial f}{\partial z}$

Solution:

(i) $\dfrac{\partial f}{\partial x} = \dfrac{\partial}{\partial x}(3x^3y - 2xy^2z) = 9x^2y - 2y^2z$

$\dfrac{\partial f}{\partial y} = \dfrac{\partial}{\partial y}(3x^3y - 2xy^2z) = 3x^3 - 4xyz$

$\dfrac{\partial f}{\partial z} = \dfrac{\partial(3x^3y - 2xy^2z)}{\partial z} = -2xy^2$

(ii) From (i),

$$\begin{aligned}\dfrac{\partial f}{\partial y} - \dfrac{\partial f}{\partial x} &= (3x^3 - 4xyz) - (9x^2y - 2y^2z)\\ &= 3x^3 - 4xyz - 9x^2y + 2y^2z\\ &= 3x^3 - 9x^2y - 4xyz + 2y^2z\\ &= 3x^2(x - 3y) - 2yz(2x - y)\end{aligned}$$

(iii) From (i),

$$\begin{aligned}\dfrac{\partial f}{\partial x} + \dfrac{\partial f}{\partial y} - \dfrac{\partial f}{\partial z} &= (9x^2y - 2y^2z) + (3x^3 - 4xyz) - (-2xy^2)\\ &= 9x^2y - 2y^2z + 3x^3 - 4xyz + 2xy^2\\ &= 9x^2y + 3x^3 + 2xy^2 - 2y^2z - 4xyz\\ &= 3x^2(3y + x) + 2y(xy - yz - 2xz).\end{aligned}$$

Exercise:
Given that $R = 2xy^2z^2 + 3xz^2 + 4xyz$. Find:
(i) $\dfrac{\partial R}{\partial x}$ (ii) $\dfrac{\partial R}{\partial x} + 2\dfrac{\partial R}{\partial z}$ (iii) $3\dfrac{\partial R}{\partial y} - 5\dfrac{\partial R}{\partial z}$

Integration

We have different types. We shall limit ourselves to the one's relevant to this book.

Standard Integral Table

(i) $\displaystyle\int x^n dx = \dfrac{x^{n+1}}{n+1} + k$ where k is an arbitrary constant and $n \neq -1$

(ii) $\displaystyle\int ax^{n+1} dx = a \int x^{n+1} dx = \dfrac{a(x^{n+2})}{n+2} + k; \; n \neq -2;$

(iii) $\displaystyle\int \sin x\, dx = -\cos x + k$

(iv) $\displaystyle\int \cos x\, dx = \sin x + k$

(v) $\displaystyle\int \dfrac{1}{x} dx = \ln x + k$

(vi) $\displaystyle\int b\, dx = b \int dx = bx + k$ (b, k are constants).

Type I: Direct Integration

Example: Integrate the following functions

(i) $\displaystyle\int 2x^n dx$ (ii) $\displaystyle\int (3x^2 + 2x - 4) dx$

(iii) $\displaystyle\int (4x+3)(x-2) dx$ (iv) $\displaystyle\int \dfrac{4x^3 + 2x^2 + 3}{4x} dx$

Solution

(i) $\int 2x^n dx = 2\int x^n dx = 2\left[\dfrac{x^{n+1}}{n+1}\right] = \dfrac{2}{n+1}x^{n+1} + k \quad (n \neq -1)$

(ii) $\int (3x^2 + 2x - 4)dx = 3\left(\dfrac{x^3}{3}\right) + 2\left(\dfrac{x^2}{2}\right) - 4x + k = x^3 + x^2 - 4x + k$

(iii) $\int (4x+3)(x-2)dx = \int (4x^2 - 5x - 6)dx = 4\left(\dfrac{x^3}{3}\right) - 5\left(\dfrac{x^2}{2}\right) - 6x + k$

$\qquad\qquad\qquad\qquad\;\; = \dfrac{4x^3}{3} - \dfrac{5x^2}{2} - 6x + k$

(iv) $\int \dfrac{4x^3 + 2x^2 + 3}{4x}dx = \int \left(\dfrac{4x^3}{4x} + \dfrac{2x^2}{4x} + \dfrac{3}{4x}\right)dx$

$\qquad\qquad\qquad\quad = \int \left[x^2 + \dfrac{1}{2}x + \dfrac{3}{4}\left(\dfrac{1}{x}\right)\right]dx$

$\qquad\qquad\qquad\quad = \dfrac{x^3}{3} + \dfrac{1}{4}x^2 + \dfrac{3}{4}\ln x + k$

$\qquad\qquad\qquad\quad = \dfrac{1}{3}x^3 + \dfrac{1}{4}x^2 + \dfrac{3}{4}\ln x + k$

Exercise:

Integrate the following:

(i) $\int 6dx$

(ii) $\int (3x^2 + 3x + 3)dx$

(iii) $\int (x-5)(2x-4)dx$

(iv) $\int \dfrac{(ax^3 + dx^2 + cx - r)}{2x^2}dx$

(v) $\int \dfrac{(2x^4 + 3x^2 + 4x)}{3x^2}dx$

Integration Using Algebraic Substitution

Example:
Integrate the following functions:

(i) $\int \dfrac{2}{(3x+2)^4} dx$ (ii) $\int xe^{x^2+3} dx$ (iii) $\int t^2 e^{t^3} dt$

(iv) $\int x^4(2+x^5)^3 dx$ (v) $\int \dfrac{5}{5x-3} dx$

Solution:

(i) Given $\int \dfrac{2}{(3x+2)^4} dx$

Put $u = 3x+2$, $\dfrac{du}{dx} = 3 \Rightarrow du = 3dx \Rightarrow dx = \dfrac{du}{3}$

$$\int \dfrac{2}{(3x+2)^4} = \int \dfrac{2}{u^4}\left(\dfrac{du}{3}\right) = \int \dfrac{2}{3}\dfrac{du}{u^4} = \dfrac{2}{3}\int u^{-4} du$$

$$= \dfrac{2}{3}\left(-\dfrac{1}{3}\right) u^{-3}$$

$$= \dfrac{-2}{9} u^{-3} = \dfrac{-2}{9}(3x+2)^{-3}$$

$$= \dfrac{-2}{9(3x+2)^3} + k$$

(ii) Given $\int xe^{x^2+3} dx$.

Let $p = x^2+3 \Rightarrow \dfrac{dp}{dx} = 2x \Rightarrow dp = 2xdx \Rightarrow xdx = \dfrac{dp}{2}$

$\therefore \int xe^{x^2+3} dx = \int e^p \dfrac{dp}{2} = \dfrac{1}{2}\int e^p dp = \dfrac{1}{2}e^p = \dfrac{1}{2}e^{x^2+3} + k.$

(iii) Given $\int t^2 e^{t^3} dt$.

$$\text{Put } p = t^3 \quad \therefore \quad dp = 3t^2 dt \Rightarrow t^2 dt = \frac{dp}{3}$$

$$\int t^2 e^{t^3} dt = \int \frac{1}{3} e^p dp = \frac{1}{3} e^p + k$$
$$= \frac{1}{3} e^{t^3} + k.$$

(iv) Given $\int x^4 (2 + x^5)^3 dx$.

$$\text{Put } u = 2 + x^5 \quad \therefore \quad du = 5x^4 dx \Rightarrow x^4 dx = \frac{du}{5}$$

$$\therefore \int x^4 (2+x^5)^3 dx \int u^3 \frac{du}{5} = \int \frac{1}{5} \int u^3 du = \frac{1}{20} u^4 = \frac{1}{20} (2+x^5)^4 + k.$$

(v) Given $\int \frac{5}{5x-3} dx$

Let $r = 5x - 3 \Rightarrow dr = 5dx$

$$\therefore \int \frac{5}{5x-3} dx = \int \frac{1}{5x-3} (5dx) = \int \frac{1}{r} dr = \log r + k = \log(5x+3) + k.$$

Exercise

Find the following:

(i) $\int \frac{7}{(2x+6)^3} dx$ (ii) $\int x^4 e^{x^5+5} dx$

(iii) $\int x^2 (3x^3 + 4) dx$ (iv) $\int \frac{8}{8x-3} dx$

(v) $\int (3x-4)^7 dx$

Type 3
Integration by Parts

This states that if u and v are different functions of u and v,
$\int v du = uv - \int u dv.$

Chapter 1

Example
Integrate the following:
(i) $\int x^2 e^x dx$ (ii) $\int 3x \log x\, dx$.

$$\begin{aligned}
(i)\quad \int x^2 e^x dx &= \int x^2 d(e^x) = x^2 e^x - \int e^x d(x^2) \\
&= x^2 e^x - 2\int x e^x dx \\
&= x^2 e^x - 2\int x d(e^x) \\
&= x^2 e^x - 2\left[x e^x - \int e^x dx\right] \\
&= x^2 e^x - 2x e^x + 2\int e^x dx \\
&= x^2 - 2x e^x + 2e^x + k
\end{aligned}$$

$$(ii)\quad \int 3x \log x\, dx = 3\int x \log x\, dx$$

$$\begin{aligned}
3\int x \log x\, dx &= 3\left[\frac{x^2}{2}\log x - \int \frac{x^2}{2}\frac{dx}{x}\right] \\
&= 3\left[\frac{1}{2}x^2 \log x - \frac{1}{2}\int x\, dx\right] \\
&= 3\left(\frac{1}{2}x^2 \log x - \frac{1}{4}x^2\right) + k.
\end{aligned}$$

Example:
Integrate the following:
(i) $\int \frac{x \log x}{2} dx$ (ii) $\int 4t e^t dx$.

All the integrations done above are indefinite. For indefinite integrals, arbitrary constant should be added at the end of the integration.

Chapter 1

Definite Integral

If the indefinite integral of $f(x)$ is $F(x)$, then the definite integral of $f(x)$ is given by $F(x)|_b^a = F(a) - F(b)$ where a and b are the lower limit and upper limit of $F(x)$ respectively.

Example

Find the value of $\int_1^3 (3x^2 + 3)dx$.

Here, $f(x) = 3x^2 + 3$, $F(x) = x^3 + 3x$, where $F(x) = \int f(x)dx$

$$\begin{aligned} F(x)|_1^3 = (x^3 + 3x)|_1^3 &= F(3) - F(1) \\ &= [3^3 + 3(3)] - (1+3) \\ &= (27+9) - 4 \\ &= 36 - 4 = 32. \end{aligned}$$

$$\therefore \int_1^3 (3x^2 + 3)dx = F(3) - F(1) = 36 - 4 = 32$$

Further Examples:

Evaluate the following:

(i) $\int_1^2 y^2 dy$ (ii) $\int_2^5 3dx$ (iii) $\int_0^3 (3x^2 + 2x - 1)dx$

Solution:

(i) $\int_1^2 y^2 dy = \left(\frac{1}{3}y^3\right)\Big|_1^2 = \frac{1}{3}(2^3) - \frac{1}{3}(1^3)$

$$= \frac{8}{3} - \frac{1}{3} = \frac{7}{3} = 2\frac{1}{3}$$

(ii) $\int_2^5 3dx = [3x]_2^5 = 3(5) - 3(2) = 15 - 6 = 9$

(iii) $\int_0^3 (3x^2 + 2x - 1)dx = \left[x^3 + x^2 - x\right]_0^3 = 3^3 + 3^2 - 3 = 27 + 9 - 3 = 36 - 3 = 33$

Chapter 1   17

Exercise:
Find the value of the following integral:

(i) $\displaystyle\int_1^2 2x^5 dx$ (ii) $\displaystyle\int_0^2 (2x^2 - x - 1)dx$ (iii) $\displaystyle\int_1^3 \left(\frac{x^3 - x^2}{x^2}\right) dx$

(iv) $\displaystyle\int_{-1}^2 x(4x^2)^3 dx$

Double Integral

This is an integral of the form $\displaystyle\int_a^b \int_c^d f(x,y) dx dy$

Example
Evaluate the following:

(i) $\displaystyle\int_0^3 \int_1^2 (x+y) dx dy$ (ii) $\displaystyle\int_1^3 \int_2^4 4 dx dy$ (iii) $\displaystyle\int_1^2 \int_0^1 (x - xy) dy dx$

Solution

(i) $\displaystyle\int_0^3 \int_1^2 (x+y) dx dy = \int_0^3 \left[\int_1^2 (x+y) dx\right] dy$

$$= \int_0^3 \left(\frac{x^2}{2} + xy\Big|_1^2\right) dy = \int_0^3 \left[\left(\frac{4}{2} + 2y\right) - \left(\frac{1}{2} + y\right)\right] dy$$

$$= \int_0^3 (2 + 2y) - \left(\frac{1}{2} + y\right) dy$$

$$= \int_0^3 \left[\frac{3}{2} + y\right] dy = \left[\frac{3}{2}y + \frac{y^2}{2}\right]_0^3$$

$$= \left(\frac{3}{2}(3) + \frac{3^2}{2}\right) - 0$$

$$= \left(\frac{9}{2} + \frac{9}{2}\right) = 9$$

(ii) $\displaystyle\int_1^3\int_2^4 4dxdy = \int_1^3 (4x|_2^4)dy$

$\displaystyle = \int_1^3 [4(4) - 4(2)]dy$

$\displaystyle = \int_1^3 8dy = 8y|_1^3 = 8(3) - 8(1) = 24 - 8 = 16$

(iii) $\displaystyle\int_1^2\int_0^1 (x - xy)dydx = \int_1^2 \left(xy - \frac{1}{2}xy^2\right)\bigg|_0^1 dx$

$\displaystyle = \int_1^2 \left[x(1) - \frac{1}{2}x(1^2)\right] dx$

$\displaystyle = \int_1^2 \left[x - \frac{1}{2}x\right] dx = \int_1^2 \frac{1}{2}xdx = \frac{1}{4}x^2\bigg|_1^2 = \frac{1}{4}(2^2) - \frac{1}{4}(1^2)$

$\displaystyle = 1 - \frac{1}{4} = \frac{3}{4}.$

Exercise
Find the value of the following:

(i) $\displaystyle\int_1^3\int_0^2 4x^2y^2 dydx$

(ii) $\displaystyle\int_0^2\int_0^3 2(x^2 + xy^2)dxdy$

(iii) $\displaystyle\int_1^3\int_1^2 (2x + 3y)dxdy$

(iv) $\displaystyle\int_2^5\int_1^3 dydx$

(v) $\displaystyle\int_0^2\int_0^3 (y^2 + x)dxdy$

(vi) $\displaystyle\int_1^2\int_2^3 k(3y - 4x)dydx$

Series and Statistical Combinations Used in the textbook

Series Used in the Textbook

(i) $\sum_{r=1}^{n} = 1 + 2 + 3 + 4 + \cdots + n = \frac{1}{2}n(n+1)$

(ii) $\sum_{r=1}^{n} r^2 = 1^2 + 2^2 + 3^2 + \cdots + n^2 = \frac{1}{6}n(n+1)(2n+1)$

(iii) $\sum_{r=0}^{\infty} x^r = 1 + x + x^2 + x^3 + \cdots = \frac{1}{1-x} \quad (-1 < x < 1)$

(iv) $\sum_{r=1}^{\infty} rx^{r-1} = 1 + 2x + 3x^2 + 4x^3 + \cdots = \frac{1}{(1-x)^2} \quad (-1 < x < 1)$

(v) $\sum_{x=1}^{\infty} x^2 r^{x-1} = 1 + 4r + 9r^2 + 16r^3 + \cdots = \frac{1+r}{(1-r)^3}; \quad |r| < 1$

(vi) $e^x = 1 + x + \frac{x^2}{2!} + \frac{x^3}{3!} + \cdots \quad$ (for all x)

Statistical Combinations

vii) $\binom{n}{x} = {}^nC_x = \frac{n!}{x!(n-x)!}$

(viii) $\sum_{r=0}^{n} \binom{m}{r}\binom{n}{k-r} = \binom{m+n}{k}$

Chapter 2

Random Variable

Random Variable

Definition: A random variable X is a prescription or rule by which every outcome α in the sample space S of the random experiment is assigned a number $X(\alpha)$ called the value of the random variable for the outcome α.

Note: random variable is abbreviated as $(r.v.)$.

Example:
When a fair coin is tossed twice, the sample $S = \{HH, HT, TH, TT\}$.
The outcomes are $\alpha_1 = HH$, $\alpha_2 = HT$, $\alpha_3 = TH$, $\alpha_4 = TT$.
Let X denote the number of heads appearing,

$$X(\alpha_1) = 2,\ X(\alpha_2) = 1,\ X(\alpha_3) = 1,\ X(\alpha_4) = 0.$$

From the above, the possible values of the random variable $X(\alpha)$ are $0, 1, 2$ i.e. $X(\alpha) = \{0, 1, 2\}$ or simply $X = \{0, 1, 2\}$.

Discrete Random Variable

Definition: If the set of all possible values of a random variable X is a count-

Chapter 2 21

able or countably infinite set $x_1, x_2, x_3, \ldots, x_n$ or $x_1, x_2, x_3, \ldots$, then X is said to be a discrete random variable.

Definition 2: The random variable $X(\alpha)$ is said to be discrete if it takes on finite or countably infinite number of integer values in a statistical experiment whose sample space is $S \ \forall \ \alpha \in S$.

Probability Mass Function

Definition:
If X is a discrete random variable, the function $f(x) = P(X = x)$, $x = x_1, x_2, x_3, \ldots, x_n$ that assigns probability to each possible value $x \in X$ is defined as discrete probability density function (pdf) or probability mass function (pmf).

Properties of Probability Mass Function

For all $x_i \in X$,

(i) $f(x_i) \geq 0$

(ii) $\sum_{x} f(x_i) = 1$, (for all possible values of x)

Example 1:
An unbiased coin is tossed three times. If a discrete random variable X denotes the number of heads that appears, write out the random variable with corresponding probability mass function in tabular form.

Solution: In tossing a fair coin times, the sample space
$$S = \{HHH, HHT, HTT, HTH, TTT, TTH, THH, THT\}$$
$$n(S) = 8$$

Let X represents the number of heads that appears. From the sample space S above,
$$X = \{0, 1, 2, 3\}.$$

Let $\alpha_0, \alpha_1, \alpha_2, \alpha_3$ represent the event that no head, one head, two heads and three heads appear respectively. Thus

$$\alpha_0 = \{TTT\},$$
$$\alpha_1 = \{HTT, THT, TTH\},$$
$$\alpha_2 = \{HHT, HTH, THH\}$$
$$\alpha_3 = \{HHH\}$$

From the above, $p(\alpha_0) = \frac{1}{8}$, $p(\alpha_1) = \frac{3}{8}$, $p(\alpha_2) = \frac{3}{8}$, $p(\alpha_3) = \frac{1}{8}$
Put $f(x_0) = p(\alpha_0)$, $f(x_1) = p(\alpha_1)$, $f(x_2) = p(\alpha_2)$, $f(x_3) = p(\alpha_3)$.
The required table is given below

X	0	1	2	3
$f(x)$	$\frac{1}{8}$	$\frac{3}{8}$	$\frac{3}{8}$	$\frac{1}{8}$

Example 2:
Calculate the value of a in the table below if $f(x)$ is probability mass function

x	0	1	2	3	4
$f(x)$	$\frac{1}{5}$	$2a$	a	$\frac{1}{2}$	$\frac{1}{4}$

Solution:
Since $f(x)$ is a probability mass function,

$$\sum f(x) = 1 \text{ for all } x \in X$$

$$\sum f(x) = \frac{1}{5} + 2a + a + \frac{1}{2} + \frac{1}{4} = 1$$

i.e. $\frac{1}{5} + 2a + a + \frac{1}{2} + \frac{1}{4} = 1$
$\therefore 3a + \frac{1}{5} + \frac{1}{2} + \frac{1}{4} = 1$
$\therefore 3a + \frac{19}{20} = 1$
$\therefore 3a = 1 - \frac{19}{20} = \frac{1}{20}$

Chapter 2

$$\therefore \quad a = \frac{1}{60}$$

Example 3:

Given that $f(x) = \dfrac{2x-1}{16}$ for $x = 1, 2, 3, 4$, show that $f(x)$ is a probability mass function.

Given that

$$\begin{aligned}
f(x) &= \frac{2x-1}{16} \\
f(1) &= \frac{2(1)-1}{16} = \frac{1}{16} \\
f(2) &= \frac{2(2)-1}{16} = \frac{3}{16} \\
f(3) &= \frac{2(3)-1}{16} = \frac{5}{16} \\
f(4) &= \frac{2(4)-1}{16} = \frac{7}{16}
\end{aligned}$$

From the above, for all x,

$$f(x) \geq 0 \tag{1}$$

$$\begin{aligned}
\sum_{x=1}^{4} f(x) &= f(1) + f(2) + f(3) + f(4) \\
&= \frac{1}{16} + \frac{3}{16} + \frac{5}{16} + \frac{7}{16} \\
&= \frac{16}{16} = 1
\end{aligned}$$

$$\sum_{x=1}^{4} f(x) = 1 \tag{2}$$

From (1) and (2), $f(x)$ satisfies the two conditions for a probability mass function, hence $f(x)$ is a probability mass function.

Example 4:

Is $f(x) = \dfrac{3x-2}{8}$ for $x = 1, 2, 3$ a probability mass function?

Solution:

Given that

$$f(x) = \dfrac{3x-2}{8} \text{ for } x = 1, 2, 3$$

$$f(1) = \dfrac{3(1)-2}{8} = \dfrac{1}{8}$$

$$f(2) = \dfrac{3(2)-2}{8} = \dfrac{4}{8}$$

$$f(3) = \dfrac{3(3)-2}{8} = \dfrac{7}{8}$$

From the above

$$f(x) > 0 \text{ for all } x \in X \tag{1}$$

Now,

$$\sum_{x=1}^{3} f(x) = f(1) + f(2) + f(3)$$

$$= \dfrac{1}{8} + \dfrac{4}{8} + \dfrac{7}{8}$$

$$= \dfrac{12}{8}$$

$$\therefore \sum_{x=1}^{3} f(x) = 1.5 > 1 \tag{2}$$

From (1) and (2),

For all x, $f(x) > 0$ and $\sum\limits_{x=1}^{3} f(x) > 1$.

Since $f(x)$ satisfies only one condition but not the two conditions for a probability mass function, therefore $f(x)$ is not a probability mass function.

Chapter 2

Mean and Variance of a Random Variable

Let X be a random variable with probability mass function $f(x)$, then the expected value of X is defined by

$$E(X) = \sum_x x f(x) \quad \text{(for discrete random variable)}$$

The sum is over all possible values of X.
The expected value of X is also referred to as the mean or expectation of X. It is also denoted by μ_x or μ. In other words, $E(X) = \mu_x = \mu$. (X is either discrete or continuous).

Properties of Expected Value of a Random Variable

If X is a random variable discrete or continuous with probability density function $f(x)$:

(i) $E(tX) = tE(X) = t\mu$

(ii) $E(X + t) = E(X) + t = \mu + t$

where t is an arbitrary constant.

Proof:
(i) $E(tX) = \sum_x tx f(x) = t \sum_x x f(x) = tE(X) = t\mu$ proved.

(ii) $E(X + t) = \sum_x (x + t) f(x) = \sum_x x f(x) + \sum_x t f(x)$
$= \sum_x x f(x) + t \sum_x f(x)$
$= E(X) + t \cdot 1 \qquad (f(x) \text{ is a p.d.f.})$
$= E(X) + t$
$= \mu + t$ proved.

Note: If X and Y are independent random variables,

(i) $E(X + Y) = E(X) + E(Y)$

(ii) $E(XY) = E(X) \cdot E(Y)$

Variance of a Random Variable

Definition:
If X is a random variable (either discrete or continuous) variance of X denoted by $var(X)$ is defined as $var(X) = E[(X - \mu)^2]$. Common notations for $var(X)$ are σ_x^2 or σ^2. The non-negative square root of $Var(X)$ is defined as the standard deviation of X. It is given by $\sigma = \sigma_x = \sqrt{var(X)}$. $var(X)$ can also be expressed as $E(X^2) - \mu^2$ (where $\mu = E(X)$) i.e. $var(X) = E[(X - \mu)^2] = E(X^2) - \mu^2$.

Proof:

$$\begin{aligned} E[(X - \mu)^2] &= E[X^2 - 2\mu X + \mu^2] \\ &= E(X^2) - 2\mu E(X) + E(\mu^2) \\ &= E(X^2) - 2\mu(\mu) + \mu^2 \\ \therefore E[(X - \mu)^2] &= E(X^2) - \mu^2 \quad \text{proved} \end{aligned}$$

Properties of Variance

If X is a random variable (discrete or continuous) of a p.d.f. $f(x)$

(i) $var(X + t) = var(X)$

(ii) $var(tX) = t^2 var(X)$

Chapter 2

Proof:

$$
\begin{aligned}
\text{(i)} \ var(X+t) &= E[(X+t)^2] - [E(X+t)]^2 \quad \text{(by definition)} \\
&= E[(X+t)^2] - (\mu_x + t)^2 \quad \text{(property of } E(X)) \\
&= E[X^2 + 2Xt + t^2] - (\mu_x + t)^2 \\
&= E(X^2) + 2tE(X) + E(t^2) - (\mu_x^2 + 2\mu_x t + t^2) \\
&= E(X^2) + 2\mu_x t + t^2 - \mu_x^2 - 2\mu_x t - t^2 \\
&= E(X^2) - \mu_x^2 \\
&= E(X^2) - \mu^2 \quad (\mu_x = \mu) \\
&= var(X) \quad \text{proved} \\
\text{(ii)} \ var(tX) &= E(tX)^2 - [E(tX)]^2 \quad \text{(by definition)} \\
&= E(t^2 X^2) - [tE(X)]^2 \quad \text{(property of } E(X)) \\
&= E(t^2 X^2) - t^2 [E(X)]^2 \\
&= t^2 E(X^2) - t^2 [E(X)]^2 \quad (t \text{ is an arbitrary constant}) \\
&= t^2 \{E(X^2) - [E(X)]^2\} \\
&= t^2 [E(X^2) - \mu_x^2] \\
&= t^2 [E(x^2) - \mu^2] \\
&= t^2 var(X) \quad \text{proved.}
\end{aligned}
$$

Worked Examples on Expected Value and Variance

Worked Examples On Expected Value

Example 1

A discrete random variable X has a p.m.f. $f(x)$ as shown in the table below

x	1	2	3	4
$f(x)$	$\frac{1}{3}$	$\frac{1}{4}$	t	$\frac{1}{6}$

Use the table to find (i) the value of t, (ii) Expected value of X.

Solution:

$\because f(x)$ is a probability mass function, $\sum_{x=1}^{4} f(x) = 1$

$\therefore \sum_{x=1}^{4} f(x) = f(1) + f(2) + f(3) + f(4) = 1$

$\therefore \dfrac{1}{3} + \dfrac{1}{4} + t + \dfrac{1}{6} = 1$

$\therefore t + \dfrac{9}{12} = 1$

$\therefore t = 1 - \dfrac{9}{12} = \dfrac{3}{12} = \dfrac{1}{4}$

Substitute $\dfrac{1}{4}$ for t in the table above, we obtain a new table:

x	1	2	3	4
$f(x)$	$\frac{1}{3}$	$\frac{1}{4}$	$\frac{1}{4}$	$\frac{1}{6}$

From the table above, the expected value of X,

$$\begin{aligned} E(X) = \sum_{x=1}^{4} x f(x) &= 1(\tfrac{1}{3}) + 2(\tfrac{1}{4}) + 3(\tfrac{1}{4}) + 4(\tfrac{1}{6}) \\ &= \tfrac{1}{3} + \tfrac{1}{2} + \tfrac{3}{4} + \tfrac{4}{6} \\ &= \tfrac{4 + 6 + 9 + 8}{12} \\ &= \tfrac{27}{12} = 2\tfrac{3}{12} = 2\tfrac{1}{4} \end{aligned}$$

$\therefore$ Expected value of $X = E(X) = 2\dfrac{1}{4}$

Example 2:

An unbiased coin is tossed twice, if X denotes the number of tails that appears,

(i) write the random variable with the corresponding probability mass function

in tabular form
(ii) find the expected value of X.

Solution:

In tossing an unbiased coin twice,

(i) the sample space $S = \{HH\ HT\ TH\ TT\}$; $n(S) = 4$
If X represents the number of tails that appears,

$$X = \{0, 1, 2\}$$

Let $\alpha_0, \alpha_1, \alpha_2$ represent the event that no tail, one tail, two tails appear respectively.

$$\alpha_0 = \{HH\},\ P(\alpha_0) = \frac{1}{4}$$
$$\alpha_1 = \{HT,\ TH\};\ P(\alpha_1) = \frac{1}{2}$$
$$\alpha_2 = \{TT\},\ P(\alpha_2) = \frac{1}{4}$$

Put $f(x_0) = P(\alpha_0), f(x_1) = P(\alpha_1)$ and $f(x_2) = P(\alpha_2)$;
Writing the information above in tabular form we obtain

x	0	1	2
$f(x)$	$\frac{1}{4}$	$\frac{1}{2}$	$\frac{1}{4}$

(ii) From the table above, the expected value of

$$X = E(X) = \sum_x x f(x)$$
$$= 0\left(\frac{1}{4}\right) + 1\left(\frac{1}{2}\right) + 2\left(\frac{1}{4}\right)$$
$$= \frac{1}{2} + \frac{1}{2} = 1$$
$$\therefore E(X) = 1.$$

Example 3:

In a gambling game, a man is paid ₦6.00 if a prime number greater than 2 appears in a single throw of a fair die. He pays out ₦2.00 if an even number occurs. Calculate his expected gain or loss.

Solution:

If a die is thrown once, the sample space $S = \{1, 2, 3, 4, 5, 6\}$; $n(S) = 6$.
Let A represent the event that a prime number greater than 2 appears

$$A = \{3, 5\}; n(A) = 2$$
$$P(A) = \frac{n(A)}{n(S)} = \frac{2}{6} = \frac{1}{3}$$

Let B represent the event that an even number appears

$$\therefore B = \{2, 4, 6\}$$

Here $n(B) = 3$, $P(B) = \dfrac{n(B)}{n(S)} = \dfrac{3}{6} = \dfrac{1}{2}$

If the man is paid ₦6.00 if a prime number greater than 2 appears and pays out ₦2.00 if an even number appears, then $X = \{6, -2\}$.
The distribution is as shown in the table below

x	6	-2
$f(x)$	$\frac{1}{3}$	$\frac{1}{2}$

$\therefore$ The expected value of

$$X = E(X) = \sum_x x f(x) = 6\left(\frac{1}{3}\right) + -2\left(\frac{1}{2}\right)$$
$$= 2 - 1$$
$$= 1.$$

$\therefore$ $E(X) = 1$, is positive, then the gambler gains ₦1.00.

Example 4

If $f(x) = k(7 - 2x); \quad x = 0, 1, 2, 3$ is a probability mass function find the value of k and the mean of X.

Solution

$\because f(x)$ is a probability mass function,

$$\sum_x f(x) = 1$$

$\therefore \sum_x k(7 - 2x) = 1$

for $x = 0, 1, 2, 3$

$k[7 - 2(0)] + k[7 - 2(1)] + k[7 - 2(2)] + k[7 - 2(3)] = 1$

$\therefore 7k + 5k + 3k + k = 1$

$\therefore 16k = 1$

$\therefore k = \dfrac{1}{16}$

$\therefore f(x) = \dfrac{1}{16}(7 - 2x), \ f(0) = \dfrac{7}{16}, \ f(1) = \dfrac{5}{16}, \ f(2) = \dfrac{3}{16}, \ f(3) = \dfrac{1}{16}$

for $k = 0, 1, 2, 3$

$$\begin{aligned}
E(X) &= \sum_x x f(x) \quad x = 0, 1, 2, 3 \\
&= 0 f(0) + 1 f(1) + 2 f(2) + 3 f(3) \\
&= 0 \left(\dfrac{7}{16}\right) + 1 \left(\dfrac{5}{16}\right) + 2 \left(\dfrac{3}{16}\right) + 3 \left(\dfrac{1}{16}\right) \\
&= 0 + \dfrac{5}{16} + \dfrac{6}{16} + \dfrac{3}{16} \\
&= \dfrac{14}{16} = \dfrac{7}{8} \\
E(X) &= \dfrac{7}{8}
\end{aligned}$$

Chapter 2

Worked Examples on Variance

Example 1
A random variable X with assigned probability is defined by

$$\begin{cases} 1 & \text{prob.} \ \frac{1}{6} \\ 2 & \text{prob.} \ \frac{2}{3} \\ 3 & \text{prob.} \ \frac{1}{6} \end{cases}$$

Use the table above to find (i) $var(X)$, (ii) $var(\frac{3}{4}X)$, (iii) $var(X+4)$.

Solution:
Presenting the above information in tabular form, we have

x	1	2	3
$f(x)$	$\frac{1}{6}$	$\frac{2}{3}$	$\frac{1}{6}$

(i) $var(X) = E(X^2) - [E(X)]^2$.
From the table,

$$\begin{aligned} E(X) = \sum_{x=1}^{3} x f(x) &= 1f(1) + 2f(2) + 3f(3) \\ &= 1\left(\frac{1}{6}\right) + 2\left(\frac{2}{3}\right) + 3\left(\frac{1}{6}\right) \\ &= \frac{1}{6} + \frac{4}{3} + \frac{3}{6} \\ &= \frac{12}{6} \\ &= 2 \end{aligned}$$

$$E(X^2) = \sum_{x=1}^{3} x^2 f(x) = 1^2 f(1) + 2^2 f(2) + 3^2 f(3)$$
$$= 1\left(\frac{1}{6}\right) + 4\left(\frac{2}{3}\right) + 9\left(\frac{1}{6}\right)$$
$$= \frac{1}{6} + \frac{8}{3} + \frac{9}{6}$$
$$= \frac{26}{6}$$
$$= \frac{13}{3} = 4\frac{1}{3}$$

$$\therefore \operatorname{var}(X) = E(X^2) - [E(X)]^2$$
$$= \frac{13}{3} - 2^2$$
$$= \frac{13}{3} - 4 = \frac{1}{3}$$
$$\therefore \operatorname{var}(X) = \frac{1}{3}$$

(ii) $\operatorname{var}\left(\frac{3}{4}X\right) = \left(\frac{3}{4}\right)^2 \operatorname{var}(X)$
$$= \frac{9}{16}\left(\frac{1}{3}\right) = \frac{3}{16}$$

(iii) $\operatorname{var}(X+4) = \operatorname{var}(X) = \frac{1}{3}$

Example 2

(i) Show that for $x = 1, 2, 3$
$$f(x) = \frac{1}{14}\binom{4}{x} \text{ is a p.m.f.}$$

(ii) Find (a) $\operatorname{var}(2X)$ (b) $\operatorname{var}(3X + 21)$

Solution:

(i) For $x = 1$,

$$\begin{aligned} f(1) &= \frac{1}{14}\binom{4}{1} \\ &= \frac{1}{14}\left(\frac{4!}{1!(4-1)!}\right) = \frac{1}{14}\left(\frac{4!}{3!}\right) = \frac{4}{14} > 0 \end{aligned}$$

For $x = 2$,

$$\begin{aligned} f(2) &= \frac{1}{14}\binom{4}{2} \\ &= \frac{1}{14}\left(\frac{4!}{2!(4-2)!}\right) = \frac{1}{14}\left(\frac{4!}{2!2!}\right) = \frac{6}{14} > 0 \end{aligned}$$

For $x = 3$,

$$\begin{aligned} f(3) &= \frac{1}{14}\binom{4}{3} \\ &= \frac{1}{14}\left(\frac{4!}{3!(4-3)!}\right) = \frac{1}{14}\left(\frac{4!}{3!}\right) = \frac{4}{14} > 0 \end{aligned}$$

From the above, we have the table below

x	1	2	3
$f(x)$	$\frac{4}{14}$	$\frac{6}{14}$	$\frac{4}{14}$

$$\begin{aligned} \therefore \sum_{x=1}^{3} f(x) &= f(1) + f(2) + f(3) \\ &= \frac{4}{14} + \frac{6}{14} + \frac{4}{14} \\ &= \frac{14}{14} \\ &= 1 \end{aligned}$$

Chapter 2

Hence, for $x = 1, 2, 3$
(i) $f(x) > 0$
(ii) $\sum_x f(x) = 1$

Since $f(x)$ satisfies the two conditions for a probability mass function $f(x)$ is a p.m.f.

(ii)(a) $Var(X) = E(X^2) - [E(X)]^2$

$$E(X) = \sum_{x=1}^{3} xf(x) = 1f(1) + 2f(2) + 3f(3)$$
$$= 1\left(\frac{4}{14}\right) + 2\left(\frac{6}{14}\right) + 3\left(\frac{4}{14}\right)$$
$$= \frac{4}{14} + \frac{12}{14} + \frac{12}{14}$$
$$= \frac{28}{14} = 2$$

$E(X) = 2$

$$E(X^2) = \sum_{x=1}^{3} x^2 f(x) = 1^2 f(1) + 2^2 f(2) + 3^2 f(3)$$
$$= \frac{4}{14} + 4\left(\frac{6}{14}\right) + 9\left(\frac{4}{14}\right)$$
$$= \frac{4}{14} + \frac{24}{14} + \frac{36}{14}$$
$$= \frac{64}{14}$$
$$= 4\frac{4}{7}$$

$$
\begin{aligned}
\therefore Var(X) &= E(X^2) - [E(X)]^2 \\
&= \frac{64}{14} - 2^2 \\
&= \frac{32}{7} - 4 \\
&= \frac{4}{7} \\
Var(X) &= \frac{4}{7} \\
Var(2X) &= 4Var(X) \\
&= 4\left(\frac{4}{7}\right) \\
&= \frac{16}{7} \\
&= 2\frac{2}{7} \\
\text{(b) } Var(3X+21) &= Var(3X) \\
&= 3^2 Var(X) \\
&= 9Var(X) \\
&= 9\left(\frac{4}{7}\right) \\
&= \frac{36}{7} = 5\frac{1}{7} \\
\therefore Var(3X+21) &= 5\frac{1}{7}.
\end{aligned}
$$

Example 3
In the table below, $f(x)$ is a p.m.f. $r + t = 1$ and $c > d$

x_i	c	d
$f(x_i)$	r	t

$i = 1, 2$

Chapter 2

Use the table to show that

(i) the expected value of $X = cr + dt$

(ii) the standard deviation of $X = (c - d)\sqrt{rt}$

Solution

$$
\begin{aligned}
\text{(i) } E(X) &= \sum x_i f(x_i) \quad (i = 1, 2) \\
&= x_1 f(x_1) + x_2 f(x_2) \\
&= cr + dt \\
E(X) &= cr + dt \quad \text{proved} \\
E(X^2) &= \sum x_i^2 f(x_i) \\
&= x_1^2 f(x_1) + x_2^2 f(x_2) \\
&= c^2 r + d^2 t \\
\therefore Var(X) &= E(X^2) - [E(X)]^2 \\
&= c^2 r + d^2 t - (cr + dt)^2 \\
&= c^2 r + d^2 t - (c^2 r^2 + 2crdt + d^2 t^2) \\
&= c^2 r + d^2 t - c^2 r^2 - 2crdt - d^2 t^2 \\
&= c^2 r - c^2 r^2 + d^2 t - d^2 t^2 - 2crdt \\
&= c^2 r(1 - r) + d^2 t(1 - t) - 2crdt \\
&= c^2 rt + d^2 tr - 2crdt \quad (r + t = 1 \text{ given}) \\
&= rt(c^2 + d^2 - 2cd) \\
&= rt(c - d)^2 \\
\text{The standard deviation} &= \sqrt{Var(X)} \\
&= \sqrt{rt(c - d)^2} = (c - d)\sqrt{rt} \quad \text{proved}
\end{aligned}
$$

Example 4

X_1 and X_2 are two stochastically independent random variables. The variance of X_1 and X_2 are $\sigma_1^2 = k$ and $\sigma_2^2 = 2$ respectively. Given that $Y = 5X_2 + X_1$ and $Var(Y) = 60$, find k.

Solution

$$\begin{aligned} Var(Y) &= Var(5X_2 + X_1) \\ 60 &= Var(5X_2) + Var(X_1) \quad (X_1, X_2 \text{ are stochastically independent}) \\ &= 5^2 Var(X_2) + Var(X_1) \\ &= 25(2) + k \\ \therefore 60 &= 50 + k \\ \therefore k &= 10 \end{aligned}$$

Cummulative Distribution Function (CDF)

Definition

The cumulative distribution function (CDF) of a random variable X is defined for any real x by $F(x) = P(X \leq x)$.

The function $F(x)$ is often referred to simply as the distribution function of X.

Relationship between $f(x)$ and $F(x)$.

If X is a discrete random variable with pdf $f(x)$ and cumulative distribution function $F(x)$; if the possible values of X are indexed in increasing order, $x_1 < x_2 < x_3 < x_4 < \cdots$ then $f(x_i) = F(x_i)$ for $i = 1$
For any $i > 1$, $f(x_i) = F(x_i) - F(x_{i-1})$.

Example:
The distribution function $F(x)$ of a random variable X is given by

$$F(x) = \begin{cases} \dfrac{2}{36}, & 0 < x < 2 \\[2mm] \dfrac{5}{36}, & 2 \leq x < 5 \\[2mm] \dfrac{1}{4}, & 5 \leq x < 10 \\[2mm] \dfrac{5}{9}, & 10 \leq x < 25 \\[2mm] 1, & x > 25 \end{cases}$$

Use the above distribution to find
(i) $P(X < 10)$ (ii) $E(X)$

Solution
Recall that if X is a discrete random variable with probability density function $f(x)$ and probability cumulative distribution function $F(x)$ and the possible values of X are indexed in increasing order $x_1 < x_2 < \ldots$ then $f(x_i) = F(x_i)$ for $i = 1$, for any $i > 1$, $f(x_i) = F(x_{i-1})$.

Using this relationship with the data given above,

$$
\begin{aligned}
f(x_1) &= F(x_1) \qquad \text{for } i = 1 \\
&= \frac{2}{36} \qquad \text{for any } i > 1 \\
f(x_2) &= F(x_2) - F(x_1) \\
&= \frac{5}{36} - \frac{2}{36} = \frac{3}{36} \\
f(x_3) &= F(x_3) - F(x_2) \\
&= \frac{1}{4} - \frac{5}{36} = \frac{4}{36} \\
f(x_4) &= F(x_4) - F(x_3) \\
&= \frac{20}{36} - \frac{1}{4} = \frac{11}{36} \\
f(x_5) &= F(x_5) - F(x_4) \\
&= 1 - \frac{20}{36} = \frac{16}{36}
\end{aligned}
$$

From the above, we obtain the following table

X	0	2	5	10	25
$f(x)$	$\frac{2}{36}$	$\frac{3}{36}$	$\frac{4}{36}$	$\frac{11}{36}$	$\frac{16}{36}$

(i) $\begin{aligned}[t] P(X < 10) &= P(X = 0) + P(X = 2) + P(X = 5) \\ &= \frac{2}{36} + \frac{3}{36} + \frac{4}{36} \\ &= \frac{9}{36} \\ &= \frac{1}{4} \end{aligned}$

Chapter 2   41

(ii) $E(X) = \sum x f(x)$
$= 0\left(\dfrac{2}{36}\right) + 2\left(\dfrac{3}{36}\right) + 5\left(\dfrac{4}{36}\right) + 10\left(\dfrac{11}{36}\right) + 25\left(\dfrac{16}{36}\right)$
$= \dfrac{6}{36} + \dfrac{20}{36} + \dfrac{110}{36} + \dfrac{400}{36}$
$= \dfrac{536}{36}$
$= 14.889$

Covariance of Two Random Variables

Definition:
Given two dependent random variables X and Y with mean μ_x and μ_y respectively, $Cov(XY)$ is defined by

$$Cov(XY) = E(XY) - \mu_x \mu_y$$

Proof:

$Cov(XY) = E(X - \mu_x)(Y - \mu_y)$
$= E(XY - X\mu_y - Y\mu_x + \mu_x\mu_y)$
$= E(XY) - \mu_y E(X) - \mu_x E(Y) + E(\mu_x\mu_y)$
$= E(XY) - \mu_y\mu_x - \mu_x\mu_y + \mu_x\mu_y$
$= E(XY) - \mu_x\mu_y - \mu_x\mu_y + \mu_x\mu_y$
$= E(XY) - \mu_x\mu_y$

If X and Y are independent, $Cov(XY) = 0$

$Cov(XY) = E(XY) - \mu_x\mu_y$
$= E(X)E(Y) - \mu_x\mu_y$
$= \mu_x\mu_y - \mu_x\mu_y$
$= 0 \quad \text{proved}$

For two dependent random variables $X_1, X_2;\ X_1 > X_2$
(i) $Var(X_1 + X_2) = Var(X_1) + Var(X_2) + 2Cov(X_1 X_2)$

(ii) $Var(X_1 - X_2) = Var(X_1) + Var(X_2) - 2Cov(X_1 X_2)$

Proof:

(i) $\begin{aligned} Var(X_1 + X_2) &= E[(X_1 + X_2) - (\mu_1 + \mu_2)]^2 \\ &= E[(X_1 - \mu_1) + (X_2 - \mu_2)]^2 \\ &= E[(X_1 - \mu_1)^2 + (X_2 - \mu_2)^2 + 2(X_1 - \mu_1)(X_2 - \mu_2)] \\ &= E(X_1 - \mu_1)^2 + E(X_2 - \mu_2)^2 + 2E(X_1 - \mu_1)(X_2 - \mu_2) \\ &= Var(X_1) + Var(X_2) + 2Cov(X_1 X_2), \quad \text{proved} \end{aligned}$

To show that $Var(X_1 - X_2) = Var(X_1) + Var(X_2) - 2Cov(X_1 X_2)$

(ii) $\begin{aligned} Var(X_1 - X_2) &= E[(X - X_2) - (\mu_1 - \mu_2)]^2 \\ &= E[(X_2 - \mu_1) - (X_2 - \mu_2)]^2 \\ &= E[(X_1 - \mu_1)^2 + (X_2 - \mu_2)^2 - 2(X_1 - \mu_1)(X_2 - \mu_2)] \\ &= E(X_1 - \mu_1)^2 + E(X_2 - \mu_2)^2 - 2E(X_1 - \mu_1)(X_2 - \mu_2) \\ &= Var(X_1) + Var(X_2) - 2Cov(X_1 X_2) \quad \text{proved.} \end{aligned}$

(iii) For two independent random variables $X_1, X_2;\ X_1 > X_2$,
$Var(X_1 + X_2) = Var(X_1 - X_2)$.

Proof:
By definition

$\begin{aligned} Var(X_1 + X_2) &= Var(X_1) + Var(X_2) + 2Cov(X_1 X_2) \\ &= Var(X_1) + Var(X_2) + 2(0) \quad (X_1, X_2 \text{ are independent}) \\ &= Var(X_1) + Var(X_2) \quad \ldots (1) \\ Var(X_1 - X_2) &= Var(X_1) + Var(X_2) - 2Cov(X_1 X_2) \\ &= Var(X_1) + Var(X_2) - 2(0) \quad (X_1, X_2 \text{ are independent}) \\ &= Var(X_1) + Var(X_2) \quad \ldots (2) \end{aligned}$

From (1) and (2),

$$Var(X_1 + X_2) = Var(X_1 - X_2) \quad \text{proved.}$$

Chapter 3

Mean and Variance of Some Discrete Probability Distributions

Bernoulli Distribution

This is a distribution with two outcomes classified into true or false, yes or no, success or failure. The probability of success is p and the probability of failure is $q = 1 - p$.

It is a distribution with probability density function (p.d.f.):

$$f(x) = \begin{cases} p^x q^{1-x} & , x = 0, 1 \\ 0 & , \text{elsewhere} \end{cases}$$

and $p + q = 1$.

The mean of Bernoulli distribution is p and its variance is pq. Symbolically $\mu = p$ and $\sigma^2 = pq$.

Proof

If X is a random variable with Bernoulli distribution.

$$\therefore \text{ Mean of } X = E(X) = \sum_{i=1}^{2} x_i f(x_i)$$
$$= x_1 f(x_1) + x_2 f(x_2)$$
$$= x_1 p^{x_1} q^{1-x_1} + x_2 p^{x_2} q^{1-x_2}$$
$$= 0.p^0 \cdot q^{1-0} + 1p^1 q^{1-1} \text{ (for } x_1 = 0, x_2 = 1\text{)}$$
$$= 0 + pq^0$$
$$= p \text{ proved.}$$

Hence the mean (μ) of a Bernoulli distribution is p.

$$E(X^2) = \sum_{i=1}^{2} x_i^2 f(x_i)$$
$$= x_1^2 f(x_1) + x_2^2 f(x_2)$$
$$= x_1^2 p^{x_1} q^{1-x_1} + x_2^2 p^{x_2} q^{1-x_2}$$
$$= 0^2 p^0 q^{1-0} + 1^2 p^1 q^{1-1}$$
$$= 0 + pq^0$$
$$= p$$
$$\therefore \text{ Variance } = E(X^2) - [E(X)]^2$$
$$= p - p^2$$
$$= p(1-p)$$
$$= pq \text{ proved.}$$

Hence the variance (σ^2) of a Bernoulli distribution is pq.

Binomial Distribution

This distribution deals with repeated and independent trials of an experiment with two outcomes resulting in either success or failure, 0 or 1, true or false, yes or no.

If the interest is on the number of successes and not in the order in which they occur, the probability of exactly x successes in n repeated trials is given

Chapter 3

by

$$f(x) = \begin{cases} \binom{n}{x} p^x q^{n-x}; & x = 0, 1, 2, \ldots, n \\ 0; & \text{elsewhere} \end{cases}$$

where p is the probability of success and $q = 1 - p$ is the probability of failure, x is the number of successes in repeated trials and $f(x)$ is the probability density function of a random variable X with binomial distribution.

Note: $\binom{n}{x} = {}^nC_x = \dfrac{n!}{x!(n-x)!}$.

Properties of Binomial Distribution

(i) It has n independent trials

(ii) It has constant probability of success (p) and probability of failure $q = 1 - p$.

(iii) There is assigned probability to non-occurrence of events.

(iv) Each trial can result in one of only two possible outcomes called success or failure.

The Mean and Variance of a Binomial Distribution

The mean (μ) of a Binomial distribution is given by np. Its variance (σ^2) is npq.

Chapter 3

Proof
Let X represent a random variable with Binomial distribution

$$\begin{aligned}
\text{Mean of } X = E(X) &= \sum x f(x) \\
&= \sum_{x=0}^{n} x \binom{n}{x} p^x q^{n-x} \quad x = 0, 1, 2, \ldots, n \\
&= \sum_{x=0}^{n} \frac{xn!}{x!(n-x)!} p^x q^{n-x} \\
&= np \sum_{x=1}^{n} \frac{(n-1)!}{(x-1)!(n-x)!} p^{x-1} q^{n-x}
\end{aligned}$$

Put $t = x - 1$,

$$\begin{aligned}
\therefore E(X) &= np \sum_{t=0}^{n-1} \frac{(n-1)!}{t!(n-t-1)!} p^t q^{n-1-t} \\
&= np \sum_{t=0}^{n-1} \binom{n-1}{t} p^t q^{n-1-t} \\
&= np(p+q)^{n-1} = np \\
&= np \text{ proved.}
\end{aligned}$$

Hence the mean (μ) of a Binomial distribution is np.

(ii) We know that $E(X^2) = E[X(X-1)] + E(X)$.

$$\begin{aligned}
\text{Now } E[X(X-1)] &= \sum_{x=0}^{n} x(x-1) \binom{n}{x} p^x q^{n-x} \\
&= \sum_{x=0}^{n} x(x-1) \left[\frac{n!}{x!(n-x)!}\right] p^x q^{n-x} \\
&= \sum_{x=2}^{n} \frac{n!}{(x-2)!(n-x)!} p^x q^{n-x} \\
&= n(n-1)p^2 \sum_{x=2}^{n} \frac{(n-2)!}{(x-2)!(n-x)!} p^{x-2} q^{n-x} \\
&= n(n-1)p^2 \sum_{x=2}^{n} \binom{n-2}{x-2} p^{x-2} q^{n-x}
\end{aligned}$$

Put $y = x - 2$ and $m = n - 2$

$$\begin{aligned}
\therefore E[X(X-1)] &= n(n-1)p^2 \sum_{y=0}^{m} \binom{m}{y} p^y q^{m-y} = n(n-1)p^2(p+q)^m \\
&= n(n-1)p^2(1) = n(n-1)p^2 \\
\text{since } E(X^2) &= E[X(X-1)] + E(X) \\
&= n(n-1)p^2 + np \\
\therefore Var(X) &= E(X^2) - [E(X)]^2 \\
&= n(n-1)p^2 + np - (np)^2 \\
&= n^2 p^2 - np^2 + np - n^2 p^2 \\
&= np - np^2 \\
&= np(1-p) \\
&= npq \quad \text{proved.}
\end{aligned}$$

$\therefore$ Variance of Binomial Distribution is npq.

Poisson Distribution

A random variable X has a poisson distribution if its probability density function is given by

$$f(x) = \begin{cases} \dfrac{\lambda^x e^{-\lambda}}{x!} & \text{for } x = 0, 1, 2, \ldots \\ 0 & \text{elsewhere} \end{cases}$$

The mean and variance of poisson distribution are respectively given by (i) $\mu = \lambda$ and (ii) $\sigma^2 = \lambda$.

Proof

$$\text{The mean } (\mu) = E(X) = \sum_{x=0}^{\infty} x f(x)$$

$$= \sum_{x=0}^{\infty} \frac{x \lambda^x e^{-\lambda}}{x!}$$

$$= \sum_{x=1}^{\infty} \frac{x \lambda^x e^{-\lambda}}{x!} \quad (\because \text{ for } x = 0, E(X) = 0)$$

$$= \sum_{x=1}^{\infty} \frac{x \lambda^x e^{-\lambda}}{x(x-1)!}$$

$$= \sum_{x=1}^{\infty} \frac{\lambda^x e^{-\lambda}}{(x-1)!}$$

$$= \lambda \sum_{x=1}^{\infty} \frac{\lambda^{x-1} e^{-\lambda}}{(x-1)!}$$

Chapter 3

Put $t = x - 1$

$$\therefore E(X) = \lambda \sum_{t=0}^{\infty} \frac{\lambda^t e^{-\lambda}}{t!}$$

$$= \lambda e^{-\lambda} \sum_{t=0}^{\infty} \frac{\lambda^t}{t!}$$

$$= \lambda e^{-\lambda} \left[1 + \lambda + \frac{\lambda^2}{2!} + \frac{\lambda^3}{3!} + \cdots + \frac{\lambda^r}{r!} + \cdots \right]$$

$$= \lambda e^{-\lambda} e^{\lambda} = \lambda$$

$\therefore$ Mean $E(X) = \lambda$ proved.

Hence the mean of Poisson distribution $= \lambda$.

(ii) We want to show that the variance of Poisson distribution $\sigma^2 = \lambda$.

Proof

From the relationship $E[X(X-1)] + E(X) = E(X^2)$,

$$E[X(X-1)] = \sum_{x=0}^{\infty} x(x-1) f(x)$$

$$= \sum_{x=0}^{\infty} \frac{x(x-1) \lambda^x e^{-\lambda}}{x!}$$

$$= \lambda^2 \sum_{x=2}^{\infty} \frac{\lambda^{x-2} e^{-\lambda}}{(x-2)!}$$

Put $t = x - 2$

$$\therefore E[X(X-1)] = \lambda^2 \sum_{t=0}^{\infty} \frac{\lambda^t e^{-\lambda}}{t!}$$

$$= \lambda^2 e^{-\lambda} \sum_{t=0}^{\infty} \frac{\lambda^t}{t!}$$

$$= \lambda^2 e^{-\lambda} e^{\lambda}$$

$$= \lambda^2$$

$$\begin{aligned}
\text{since } E(X^2) &= E[X(X-1)] + E(X) \\
\therefore E(X^2) &= \lambda^2 + \lambda \\
\therefore \text{Variance} &= E(X^2) - [E(X)]^2 \\
&= \lambda^2 + \lambda - \lambda^2 \\
&= \lambda.
\end{aligned}$$

$\therefore$ Variance of Poisson distribution $(\sigma^2) = \lambda$.
From the above, $\sigma^2 = \mu = \lambda$ i.e., Mean and variance of Poisson Distribution are the same.

Discrete Uniform Distribution
Definition

A random variable X has a discrete uniform distribution if and only if its probability density function is given by $f(x) = \dfrac{1}{k}$ for $x = 1, 2, 3, \ldots, k$.
The mean and variance of a discrete uniform distribution is

(i) $\mu = \dfrac{k+1}{2}$ and its variance is

(ii) $\sigma^2 = \dfrac{k^2 - 1}{12}$

Proof

$$\begin{aligned}
\text{Mean } (\mu) = E(X) &= \sum_{x=1}^{k} x f(x) \\
&= \sum_{x=1}^{k} x(1/k) \\
&= \frac{1}{k} \sum_{x=1}^{k} x
\end{aligned}$$

Recall that

$$\sum_{r=1}^{k} r = \frac{k(k+1)}{2}$$

$$\therefore E(X) = \frac{1}{k}\sum_{x=1}^{k} x = \frac{1}{k}\left(\frac{k(k+1)}{2}\right)$$

$$= \frac{k+1}{2} \quad \text{proved.}$$

(ii) We want to show that the variance $(\sigma^2) = \dfrac{k^2 - 1}{2}$.

Proof

$$\begin{aligned}
E(X^2) &= \sum_{x=1}^{k} x^2 f(x) \\
&= \sum_{x=1}^{k} x^2 \left(\frac{1}{k}\right) \\
&= \frac{1}{k}\sum_{x=1}^{k} x^2 \\
&= \frac{1}{k}[1 + 2^2 + 3^2 + \cdots + k^2] \\
&= \frac{1}{k}\left[\frac{1}{6}k(k+1)(2k+1)\right] \\
&= \frac{1}{6}(k+1)(2k+1)
\end{aligned}$$

$$\begin{aligned}
\therefore Var(X) &= E(X^2) - [E(x)]^2 \\
&= \frac{1}{6}(k+1)(2k+1) - \left(\frac{k+1}{2}\right)^2 \\
&= \frac{(k+1)(2k+1)}{6} - \frac{k^2 + 2k + 1}{4} \\
&= \frac{2(2k^2 + 3k + 1) - 3(k^2 + 2k + 1)}{12} \\
&= \frac{k^2 - 1}{12} \quad \text{proved.}
\end{aligned}$$

Hence the variance of a discrete uniform distribution $\sigma^2 = \dfrac{k^2 - 1}{12}$.

Geometric Distribution
Definition

A random variable X has geometric distribution if and only if its probability density function is given by

$$f(x) = \begin{cases} \theta(1-\theta)^{x-1} &, \quad x = 1, 2, 3, \ldots \\ 0 &, \quad \text{elsewhere} \end{cases}$$

The mean and variance of a Geometric distribution are respectively $\dfrac{1}{\theta}$ and $\sigma^2 = \dfrac{1-\theta}{\theta^2}$.

Proof

(i) $\mu = E(X) = \displaystyle\sum_{x=1}^{\infty} x\theta(1-\theta)^{x-1}$

Put $p = \theta$ and $(1-\theta) = q$

$$\begin{aligned}
\mu = E(X) &= \sum_{x=1}^{\infty} xpq^{x-1} \quad x = 1, 2, 3, \ldots \\
&= p \sum_{x=1}^{\infty} xq^{x-1} \\
&= p(1 + 2q + 3q^2 + 4q^3 + \cdots) \\
&= p\left[\dfrac{1}{(1-q)^2}\right] \\
&= p\left(\dfrac{1}{p^2}\right) = \dfrac{1}{p} \quad \because (p+q=1)
\end{aligned}$$

But $p = \theta$ (given) $\therefore E(X) = \dfrac{1}{\theta}$

Hence the mean of a geometric distribution $\mu = \dfrac{1}{\theta}$.

Chapter 3

We want to show that the variance $\sigma^2 = \dfrac{1-\theta}{\theta^2}$.

Now, $E(X^2) = \displaystyle\sum_{x=1}^{\infty} x^2 f(x)$

$\phantom{\text{Now, } E(X^2)} = \displaystyle\sum_{x=1}^{\infty} x^2 \theta(1-\theta)^{x-1}$

Put $p = \theta$ and $q = 1 - \theta$. Therefore,

$$E(X^2) = \sum_{x=1}^{\infty} x^2 p q^{x-1}$$

$$\therefore\ E(X^2) = p \sum_{x=1}^{\infty} x^2 q^{x-1}$$

$$= p[1 + 4q + 9q^2 + 16q^3 + \cdots]$$

$$= p\left(\dfrac{1+q}{(1-q)^3}\right)$$

$$= \dfrac{p(1+q)}{p^3} = \dfrac{1+q}{p^2}$$

Since $p = \theta$ and $q = 1 - \theta$ (given)

$$\therefore\ E(X^2) = \dfrac{1 + (1-\theta)}{\theta^2} = \dfrac{2-\theta}{\theta^2}$$

Variance $(\sigma^2) = E(X^2) - [E(X)]^2$

$$= \dfrac{2-\theta}{\theta^2} - \dfrac{1}{\theta^2} = \dfrac{2-\theta-1}{\theta^2} = \dfrac{1-\theta}{\theta^2}$$

Hence variance of a geometric distribution $\sigma^2 = \dfrac{1-\theta}{\theta^2}$.

Chapter 3

Alternative Method

Mean of X of geometric distribution is

$$\begin{aligned}
E(X) = \sum_{x=1}^{\infty} xpq^{x-1} &= \sum_{x=1}^{\infty} p\frac{d}{dq}q^x \quad \text{where } p = \theta \text{ and } q = 1-\theta \\
&= p\frac{d}{dq}\sum_{x=1}^{\infty} q^x \\
&= p\frac{d}{dq}(q + q^2 + q^3 + \cdots) \\
&= p\frac{d}{dq}q(1 + q + q^2 + q^3 + \cdots) \\
&= p\frac{d}{dq}q\left(\frac{1}{1-q}\right) \\
&= p\frac{d}{dq}\left(\frac{q}{1-q}\right) \\
&= p\left[\frac{(1-q)(1) - q(-1)}{(1-q)^2}\right] \\
&= p\frac{(1-q+q)}{(1-q)^2} \\
&= p\left(\frac{1}{p^2}\right) \\
&= \frac{1}{p}
\end{aligned}$$

Since $p = \theta$, $\quad \therefore E(X) = \dfrac{1}{\theta}$.

$$
\begin{aligned}
\text{Now, } E(X^2) &= \sum_{x=1}^{\infty} x^2 p q^{x-1} \\
&= p \sum_{x=1}^{\infty} \frac{d}{dq}(xq^x) \\
&= p \frac{d}{dq} \sum_{x=1}^{\infty} (xq^x) \\
&= p \frac{d}{dq}(q + 2q^2 + 3q^3 + 4q^4 + \cdots) \\
&= p \frac{d}{dq} q(1 + 2q + 3q^2 + 4q^3 + \cdots) \\
&= p \frac{d}{dq} q \left[\frac{1}{(1-q)^2}\right] \\
&= p \left[\frac{(1-q)^2(1) + 2q(1-q)}{(1-q)^4}\right] \\
&= \frac{p[(1-q)][(1-q+2q)]}{(1-q)^4} \\
&= p \left[\frac{1+q}{(1-q)^3}\right] \\
&= p \left[\frac{(1+q)}{p^3}\right] \quad (\because p = 1-q) \\
&= \frac{1+q}{p^2}
\end{aligned}
$$

Since $p = \theta$ and $q = 1 - \theta$, (given)

$$\therefore E(X^2) = \frac{1+q}{p^2} = \frac{1+(1-\theta)}{\theta^2} = \frac{2-\theta}{\theta^2}$$

$$= \frac{2-\theta}{\theta^2}$$

$$\text{Variance} = E(X^2) - [E(X)]^2$$

$$= \frac{2-\theta}{\theta^2} - \left(\frac{1}{\theta}\right)^2$$

$$= \frac{2-\theta-1}{\theta^2}$$

$$= \frac{1-\theta}{\theta^2} \quad \text{proved.}$$

Hypergeometric Distribution

A random variable X has a hypergeometric distribution if its probability distribution is given by

$$f(x, n, N, k) = \frac{\binom{k}{x}\binom{N-k}{n-x}}{\binom{N}{n}} \quad \text{for } x = 0, 1, 2, \ldots, n, \ x \leq k$$

and $n - x \leq N - k$

The mean and the variance of hypergeometric distribution are respectively
$\mu = \dfrac{nk}{N}$ and $\sigma^2 = \dfrac{nk(N-k)(N-n)}{N^2(N-1)}$.

Reqd: (i) We want to show that $\mu = \dfrac{nk}{N}$.

Chapter 3

Proof

$$\begin{aligned}
\text{Mean } (\mu) &= E(X) \\
&= \sum_{x=0}^{n} x f(x) \\
&= \sum_{x=1}^{n} x \frac{\binom{k}{x}\binom{N-k}{n-x}}{\binom{N}{n}} \\
&= \sum_{x=1}^{n} \frac{x \cdot \frac{k!}{x!(k-x)!}\binom{N-k}{n-x}}{\binom{N}{n}} \\
&= \sum_{x=1}^{n-1} \frac{\frac{k!}{(x-1)!(k-x)!}\binom{N-k}{n-x}}{\binom{N}{n}} \\
&= \frac{k}{\binom{N}{n}} \left[\sum_{x=1}^{n-1} \frac{(k-1)!}{(x-1)!(k-x)!}\right]\binom{N-k}{n-x}
\end{aligned}$$

Put $t = x - 1$ and $r = n - 1$

$$\therefore \text{Mean } (\mu) = \frac{k}{\binom{N}{n}} \sum_{t=0}^{r} \binom{k-1}{t}\binom{N-k}{r-t}$$

Recall that

$$\sum_{r=0}^{n} \binom{m}{r}\binom{n}{k-r} = \binom{m+n}{k}$$

Chapter 3
58

$$\therefore \mu = \frac{k}{\binom{N}{n}} \binom{N-1}{r} = \frac{k}{\binom{N}{n}} \binom{N-1}{n-1} \quad (\because r = n-1)$$

$$= \frac{k}{\frac{N!}{n!(N-n)!}} \times \frac{(N-1)!}{(n-1)!(N-n)!}$$

$$= \frac{kn!(N-n)!}{N!} \times \frac{(N-1)!}{(n-1)!(N-n)!}$$

$$= \frac{kn!(N-n)!}{N(N-1)!} \cdot \frac{(N-1)!}{(n-1)!(N-n)!}$$

$$= \frac{kn!}{N(n-1)!} = \frac{kn}{N}$$

$\therefore$ Mean $(\mu) = \dfrac{nk}{N}$ proved.

(ii) We want to show that the variance $(\sigma^2) = \dfrac{nk(N-k)(N-n)}{N^2(N-1)}$.

Proof
Recall that $E(X^2) = E[X(X-1)] + E(X)$.

$$\text{where } E[X(X-1)] = \sum x(x-1)f(x)$$

$$= \sum_{x=2}^{n-2} \frac{x(x-1)\binom{k}{x}\binom{N-k}{n-x}}{\binom{N}{n}}$$

$$= \sum_{x=2}^{n-2} \frac{x(x-1)k!}{x!(k-x)!} \frac{\binom{N-k}{n-x}}{\binom{N}{n}}$$

$$= \frac{k(k-1)}{\binom{N}{n}} \sum_{x=2}^{n-2} \frac{(k-2)!}{(x-2)!(k-x)!} \binom{N-k}{n-x}$$

Put $y = x - 2$, $m = n - 2$

$$
\begin{aligned}
E[X(X-1)] &= \frac{k(k-1)}{\binom{N}{n}} \sum_{y=0}^{m} \left[\binom{k-2}{y} \binom{N-k}{m-y} \right] \\
&= \frac{k(k-1)}{\binom{N}{n}} \binom{N-2}{m} \\
&= \frac{k(k-1)}{\binom{N}{n}} \binom{N-2}{n-2} \\
&= \frac{k(k-1)}{\frac{N!}{n!(N-n)!}} \left[\frac{(N-2)!}{(n-2)!(N-n)!} \right] \\
&= \frac{k(k-1)}{N!} n!(N-n)! \frac{(N-2)!}{(n-2)!(N-n)!} \\
&= \frac{k(k-1)}{N!} n! \frac{(N-2)!}{(n-2)!} \\
&= \frac{k(k-1)n(n-1)}{N(N-1)}
\end{aligned}
$$

$$\begin{aligned}
\sigma^2 &= E(X^2) - [E(X)]^2 \\
&= E[X(X-1)] + [E(X)] - [E(X)]^2 \\
&= \frac{k(k-1)n(n-1)}{N(N-1)} + \frac{nk}{N} - \left(\frac{nk}{N}\right)^2 \\
&= \frac{k(k-1)n(n-1)}{N(N-1)} + \frac{nk}{N} - \frac{n^2 k^2}{N^2} \\
&= \frac{Nk(k-1)n(n-1) + N(N-1)nk - n^2 k^2(N-1)}{N^2(N-1)} \\
&= \frac{nk[N(k-1)(n-1) + N(N-1) - nk(N-1)]}{N^2(N-1)} \\
&= \frac{nk[N(k-1)(n-1) + (N-1)(N-nk)]}{N^2(N-1)} \\
&= \frac{nk}{N^2(N-1)}[Nnk - Nn - Nk + N + N^2 - nNk - N + nk] \\
&= \frac{nk}{N^2(N-1)}[N^2 - Nn - Nk + nk] \\
&= \frac{nk}{N^2(N-1)}[N(N-n) - k(N-n)] \\
&= \frac{nk}{N^2(N-1)}(N-k)(N-n)
\end{aligned}$$

$\therefore$ Variance $\sigma^2 = \dfrac{nk(N-k)(N-n)}{N^2(N-1)}$ proved.

Chapter 4

Practical Application of Some Probability Distributions

Binomial Distribution

This distribution deals with repeated and independent trials of an experiment with two outcomes resulting in either success or failure, 0 or 1, true or false, yes or no etc. If the interest is in the number of successes and not in the order in which they occur, the probability of exactly x successes in n repeated trials is given by

$$f(x) = \begin{cases} \binom{n}{x} p^x q^{n-x} & ; \ x = 0, 1, 2, \ldots, n \\ 0 & ; \ \text{otherwise} \end{cases}$$

where $\binom{n}{x} = {}^nC_x = \dfrac{n!}{x!(n-x)!}$ and p is the probability of success and $q = 1 - p$ is the probability of failure and x is the number of successes in repeated trials. $f(x)$ gives the probability density function (p.d.f.) of the Binomial distribution.

Properties of Binomial Distribution

(i) It has n independent trials

(ii) It has constant probability of success (p) and probability of failure $q = 1 - p$ from trial to trial

(iii) Probability is assigned to non-occurrence of events.

(iv) The mean $(\mu) = np$ and variance $(\sigma^2) = npq$

(v) Each trial can result in one of only two possible outcomes called success or failure.

Note: The mean, variance and standard deviation of a binomial distribution are given respectively as
Mean $(\mu) = np$
Variance $(\sigma^2) = npq$
Standard deviation $= \sqrt{npq}$

Example 1
An investigation shows that out of every 8 patients treated with a new malaria vaccine 6 patients are cured. If 5 patients are treated with the vaccine, what is the probability that

(i) 1 patient is cured

(ii) exactly 3 patients are cured.

(iii) at least 3 patients are cured.

Give your answer to 3 decimal places.

Solution
$p = \dfrac{6}{8} = \dfrac{3}{4}$, $q = 1 - \dfrac{3}{4} = \dfrac{1}{4}$
$n = 5$.

By Binomial distribution

$$P(X = r) = {}^nC_r p^r q^{n-r}$$

(i) $P(X = 1) = {}^5C_1 \left(\dfrac{3}{4}\right)^1 \left(\dfrac{1}{4}\right)^{5-1}$

$\qquad\qquad\qquad = {}^5C_1 \dfrac{3}{4} \left(\dfrac{1}{4}\right)^4 = 0.0146 = 0.015$ (to 3 dec. places)

(ii) $P(X = 3) = {}^5C_3 \left(\dfrac{3}{4}\right)^3 \left(\dfrac{1}{4}\right)^{5-3} = {}^5C_3 \left(\dfrac{3}{4}\right)^3 \left(\dfrac{1}{4}\right)^2$

$\qquad\qquad\qquad = {}^5C_3 \left(\dfrac{27}{64}\right) \left(\dfrac{1}{16}\right) = 0.2637 = 0.264$ (to 3 dec. places)

(iii) $P(X \geq 3) = 1 - P(X < 3)$

$\qquad\qquad\qquad = 1 - [P(X = 0) + P(X = 1) + P(X = 2)$

$\qquad\qquad\qquad = 1 - \left[{}^5C_0 \left(\dfrac{3}{4}\right)^0 \left(\dfrac{1}{4}\right)^5 + {}^5C_1 \left(\dfrac{3}{4}\right)^1 \cdot \left(\dfrac{1}{4}\right)^4 + {}^5C_2 \left(\dfrac{3}{4}\right)^2 \cdot \left(\dfrac{1}{4}\right)^3 \right]$

$\qquad\qquad\qquad = 1 - [0.00097 + 0.0146 + 0.0879]$

$\qquad\qquad = 0.10347$

$\qquad\qquad\qquad = 0.103$ (to 3 dec. places)

Example 2
A fair die is tossed four times. Find the probability that a prime number or an odd number greater than one appears.

(i) exactly two times

(ii) at least three times.

Solution:
This is Binomial Distribution where

$$P(X = r) = {}^nC_r p^r q^{n-r}.$$

Chapter 4

When a die is thrown, the sample space $S = \{1, 2, 3, 4, 5, 6\}$.
Let E_1 represent the event that a prime number appears and E_2 the event that an odd number greater than one appears respectively.

$$E_1 = \{2, 3, 5\} \quad \therefore \quad P(E_1) = \frac{3}{6} = \frac{1}{2}$$

$$E_2 = \{3, 5\} \quad \therefore \quad P(E_2) = \frac{2}{6} = \frac{1}{3}$$

$\therefore$ Probability that a prime number or an odd number greater than one appears
$= P(E_1) + P(E_2) = \frac{1}{2} + \frac{1}{3} = \frac{5}{6}$

Probability of success $(p) = \frac{5}{6}$

Probability of failure $(q) = 1 - p = 1 - \frac{5}{6} = \frac{1}{6}$

No. of trials $(n) = 4$
No. of successes $(x) = 2$
$\therefore$ By Binomial distribution

$$
\begin{aligned}
\text{(i)} \quad P(X = 2) &= {}^4C_2 \left(\frac{5}{6}\right)^2 \left(\frac{1}{6}\right)^{4-2} \\
&= {}^4C_2 \left(\frac{5}{6}\right)^2 \left(\frac{1}{6}\right)^2 \\
&= 6 \left(\frac{25}{36}\right) \left(\frac{1}{36}\right) \\
&= \frac{25}{216}
\end{aligned}
$$

$\therefore$ Probability that a prime number or an odd number greater than one appears exactly two times $= \dfrac{25}{216}$.

(ii) Probability that a prime number or an odd number greater than one ap-

Chapter 4

pears at least three times is

$$\begin{aligned}
P(X \geq 3) &= 1 - P(X \leq 2) \\
&= 1 - \{P(X=0) + P(X=1) + P(X=2)\} \\
&= 1 - (\,^4C_0 p^0 q^4 + \,^4C_1 p^1 q^{4-1} + \,^4C_2 p^2 q^{4-2}\} \\
&= 1 - \{(\frac{1}{6})^4 + 4(\frac{5}{6})(\frac{1}{6})^3 + 6(\frac{5}{6})^2(\frac{1}{6})^2\} \\
&= 1 - \left\{\frac{1}{1296} + \frac{20}{1296} + \frac{150}{1296}\right\} \\
&= 1 - \left\{\frac{171}{1296}\right\} = \frac{1125}{1296}
\end{aligned}$$

Example 3
A coin is loaded in such a way that a tail is thrice as likely to appear as a head, if the coin is tossed five times, find the probability that

(i) exactly two tails appear

(ii) at least two tails appear

(iii) at most three tails appear.

(iv) obtain the mean and variance of the appearance of tail in a single toss.

Solution
The sample space $S = \{H, T\}$.
Since the coin is loaded such that a tail is three times as likely to appear as a head, if weight w is assigned to the appearance of a head weight $3w$ will be assigned to the appearance of a tail

$$\therefore w + 3w = 1 \text{ (property of probability)}$$
$$4w = 1 \Rightarrow w = \frac{1}{4}.$$

Hence the probability of a head appearing $P(H)$ is $\frac{1}{4}$ while the probability of a tail appearing is $P(T) = 1 - \frac{1}{4} = \frac{3}{4}$.

Probability of a tail appearing is the probability of success.

∴ $P(T) = \dfrac{3}{4}$

Probability of failure $= 1 - \dfrac{3}{4} = \dfrac{1}{4}$

Here $n = 5 \quad x = 2$.

∴ By Binomial Distribution

$$\begin{aligned} P(X = r) &= {}^nC_r p^r q^{n-r} \\ \text{(i)} \quad P(X = 2) &= {}^5C_2 p^2 q^{5-2} \\ &= {}^5C_2 \left(\dfrac{3}{4}\right)^2 \left(\dfrac{1}{4}\right)^{5-2} \\ &= 10 \left(\dfrac{9}{16}\right) \left(\dfrac{1}{64}\right) \\ &= \dfrac{45}{512} \end{aligned}$$

(ii) Probability of at least two tails appearing

$$\begin{aligned} P(X \geq 2) &= 1 - P(X < 2) \\ &= 1 - \{P(X = 0) + P(X = 1)\} \\ P(X = 0) &= {}^5C_0 p^0 q^5 \\ &= 1 \left(\dfrac{3}{4}\right)^0 \left(\dfrac{1}{4}\right)^5 \\ &= \dfrac{1}{1024} \\ P(X = 1) &= {}^5C_1 \left(\dfrac{3}{4}\right)^1 \left(\dfrac{1}{4}\right)^{5-1} \\ &= 5 \left(\dfrac{3}{4}\right) \left(\dfrac{1}{4}\right)^4 \\ &= \dfrac{15}{1024} \end{aligned}$$

$$\therefore P(X \geq 2) = 1 - P(X < 2) = 1 - \left(\frac{1}{1024} + \frac{15}{1024}\right) = 1 - \frac{16}{1024}$$
$$= 1 - \frac{1}{64} = \frac{63}{64}$$

(iii) Probability of at most 3 tails appear

$$P(X \leq 3) = P(X = 0) + P(X = 1) + P(X = 2) + P(X = 3)$$

From the solution above,

$$P(X = 0) = \frac{1}{1024}$$

$$P(X = 1) = \frac{15}{1024}$$

$$P(X = 2) = \frac{45}{512}$$

$$P(X = 3) = {}^5C_3 p^3 q^{5-3} = {}^5C_3 \left(\frac{3}{4}\right)^3 \left(\frac{1}{4}\right)^2 = 10 \left(\frac{27}{64}\right)\left(\frac{1}{16}\right) = \frac{270}{1024}$$

$$\therefore P(X \leq 3) = \frac{1}{1024} + \frac{15}{1024} + \frac{45}{512} + \frac{270}{1024}$$
$$= \frac{1}{1024} + \frac{15}{1024} + \frac{90}{1024} + \frac{270}{1024}$$
$$= \frac{1 + 15 + 90 + 270}{1024} = \frac{376}{1024}$$
$$= \frac{47}{128}$$

(iv) Mean $= np = 5 \times \frac{3}{4} = \frac{15}{4} = 3.75$

Variance $= npq = 5 \cdot \frac{3}{4} \cdot \frac{1}{4} = \frac{15}{16}$.

Poisson Distribution

The probability density function of a Poisson distribution is defined as

$$f(x) = \begin{cases} \dfrac{\lambda^x e^{-\lambda}}{x!} & x = 0, 1, 2, \ldots, \ \lambda > 0 \\ 0 & \text{elsewhere} \end{cases}$$

The mean of Poisson distribution is λ.
The variance of Poisson distribution is the same as its mean $= \lambda$.

Properties of Poisson Distribution

(i) The number of trials is very large and the probability of success is very small.

(ii) The population average or rate is known and is constant.

(iii) Probability is assigned to non-occurrence of events.

(iv) The mean (μ) and the variance (σ^2) are both equal.

Poisson Approximation to Binomial

A Binomial distribution with parameters n and p can be approximated by a Poisson distribution with parameters $\lambda = np$ when

(i) n is large ($n > 30$)

(ii) p is small ($p < 0.1$)

(iii) $np < 5$

Substituting np for λ in Poisson distribution

$$\therefore \ f(x) = \frac{\lambda^x e^{-\lambda}}{x!} = \frac{(np)^x e^{-np}}{x!}, \quad x = 0, 1, 2, \ldots$$

Example 1

A bank received on the average 4 bad cheques per day. What is the probability that on a particular day the bank will receive

Chapter 4

(i) no bad cheque,

(ii) 2 bad cheques,

(iii) 3 bad cheques,

(iv) between 2 and 4 bad cheques inclusive.

Solution
$\lambda = 4$
Recall that $P(X = x) = \dfrac{e^{-\lambda}\lambda^x}{x!}$
Therefore, $P(X = x) = \dfrac{e^{-4}4^x}{x!}$

(i) $P(X = 0) = \dfrac{e^{-4}4^0}{0!} = e^{-4} = 0.018$

(ii) $P(X = 2) = \dfrac{e^{-4}4^2}{2!} = e^{-4} \times 8 = 8e^{-4} = 0.144$

(iii) $P(X = 3) = \dfrac{e^{-4}4^3}{3!} = \dfrac{32}{3}e^{-4} = 0.018315638 \times \dfrac{32}{3} = 0.195$

(iv) $P(\text{between 2 and 4 inclusive})$
$= P(2 \leq X \leq 4) = P(X = 2) + P(X = 3) + P(X = 4)$

From the above solution

$$P(X = 2) = 0.144 \qquad P(X = 3) = 0.195$$

Now

$$\begin{aligned} P(X = 4) &= \dfrac{e^{-4}4^4}{4!} = \dfrac{256}{24}e^{-4} = \dfrac{32}{3}e^{-4} = 0.195 \\ P(2 \leq X \leq 4) &= 0.144 + 0.195 + 0.195 \\ &= 0.534 \end{aligned}$$

Example 2
In a manufacturing industry, for every 10,000 bulbs produced 5 were found defective. If 500 bulbs were produced, find the probability that:

(i) none is defective

(ii) at most 2 are defective

(iii) at least 3 are defective.

Solution
Since $n > 30$ and $p < 0.1$ we apply Poisson approximation.
Here $n = 500 \quad p = \dfrac{5}{10000} = 5 \times 10^{-4}$
$\therefore \lambda = np = 500 \times 5 \times 10^{-4} = 25 \times 10^{-2} = 0.25$
Recall that
$$P(X = x) = \frac{e^{-\lambda}\lambda^x}{x!} = \frac{e^{-0.25}(0.25)^x}{x!}$$

(i) Probability (that none is defective) $= P(X = 0) = \dfrac{e^{-\lambda}\lambda^0}{0!}$
$$= \frac{e^{-0.25}(0.25)^0}{0!} = e^{-0.25} = 0.779$$

(ii) $P(\text{at most 2 are defective}) = P(X \leq 2) = P(X = 0)$
$\qquad\qquad\qquad\qquad\qquad\qquad\qquad\qquad + P(X = 1) + P(X = 2)$
$\qquad\qquad\qquad\qquad P(X = 0) = e^{-0.25} = 0.779$
$\qquad\qquad\qquad\qquad P(X = 1) = \dfrac{e^{-0.25}(0.25)^1}{1!}$
$\qquad\qquad\qquad\qquad\qquad\qquad = e^{-0.25}(0.25) = 0.195$

$$P(X = 2) = \frac{e^{-0.25}(0.25)^2}{2!} = \frac{1}{2}(0.0625 \times e^{-0.25}) = 0.024$$
$$\therefore P(X \leq 2) = e^{-0.25} + e^{-0.25}(0.25) + \frac{1}{2}(6.25 \times 10^{-2} \times e^{-0.25})$$
$$= 0.779 + 0.195 + 0.024 = 0.998$$

$P(\text{at least 3 are defective})$
$$= P(X \geq 3) = 1 - P(X \leq 2)$$
$$= 1 - 0.998 = 0.002$$

Chapter 4

Normal Distribution

The probability density function of a normal distribution is given as

$$f(x) = \frac{1}{\sigma\sqrt{2\pi}} e^{-\frac{1}{2}(x-\mu)^2} \; ; \quad -\infty < x < \infty. \qquad (i)$$

where μ and σ are mean and standard deviation respectively.

The normal distribution with mean $\mu = 0$ and $\sigma = 1$ is called the standard normal distribution.

Hence a random variable Z is said to have a standardized normal distribution if its probability density function is given as

$$f(z) = \frac{1}{\sqrt{2\pi}} e^{-z^2/2}$$

where $Z \sim N(0,1)$ and Z is the normal variate with mean zero and variance one.

The graph of a normal distribution $f(z)$ is as shown below

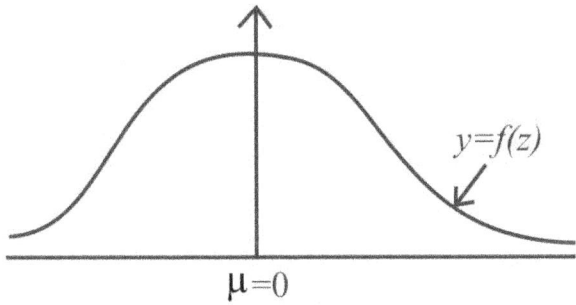

Features of the Normal Distribution

(i) The mean and standard deviation of a standardised normal distribution are zero and one respectively.

(ii) Its curve is symmetrical about a vertical axis through the mean.

(iii) The coefficient of skewness of a standardised normal distribution is zero.

(iv) The total area under the curve and above the horizontal axis is equal to one

(v) Its curve approaches the horizontal axis asymptotically as we proceed in either direction away from the mean

(vi) The Y-axis about which the curve is symmetrical divides the area enclosed by the curve and the X-axis into two equal parts each of area 0.5.

Calculation of the area under the normal curve

Points to note in calculating the area under the normal curve.

Basically symmetric property of the normal curve is used in computing the required area. The cases considered below serve as useful aid in our computation.

Case (i)

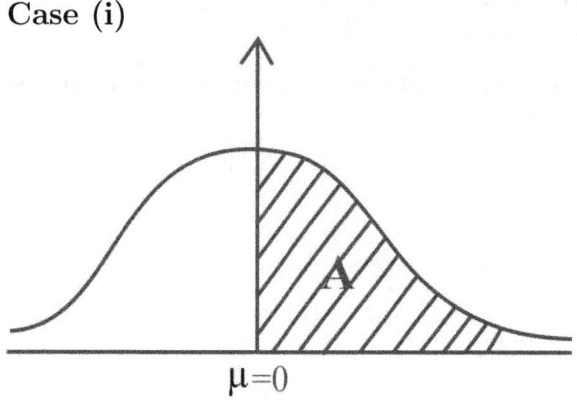

Area $A = P(Z \geq 0) = 0.5$

Chapter 4

Case (ii)

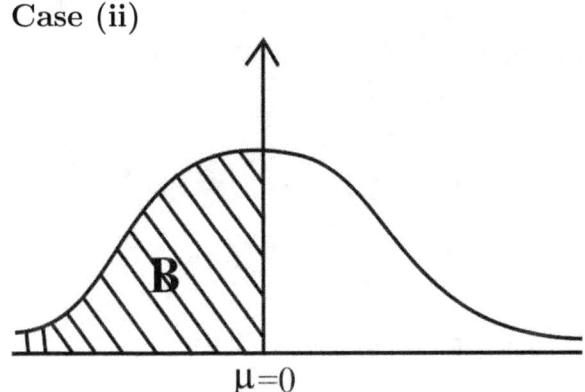

$\mu = 0$

Area $B = P(Z \leq 0) = 0.5$

Note: Case (i) and Case (ii) give the standard value of $P(Z \geq 0)$ and $P(Z \leq 0)$.

Case (iii)

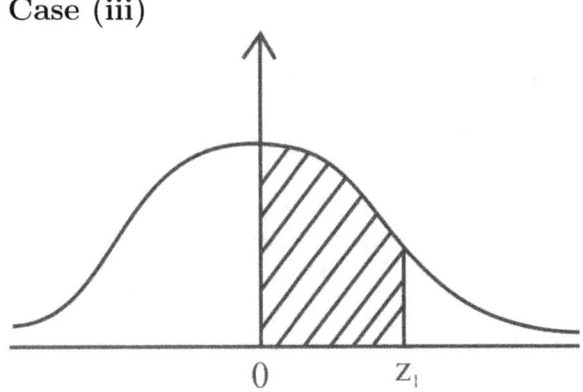

$0 \quad Z_1$

The shaded area $= P(0 < Z < Z_1)$

If $Z_1 = 1.23$, $\therefore$ the shaded area $= P(0 < Z < 1.23) = 0.3907$

Note: Only the values of $P(0 < Z < Z_1)$ are shown on the normal table.

Case (iv)

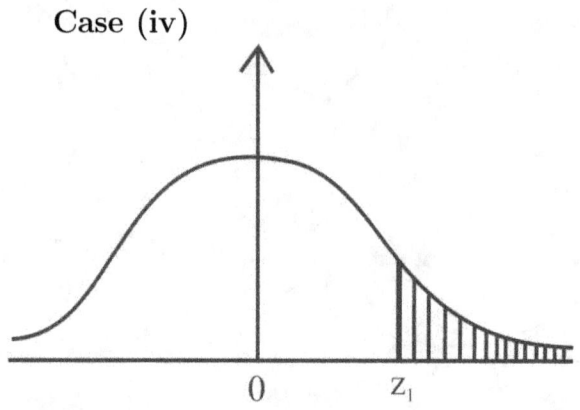

$$\text{Area of } A = P(Z \geq Z_1)$$
$$= P(Z \geq 0) - P(0 < Z < Z_1)$$

If $Z_1 = 2.1$

$$\therefore P(Z \geq 2.1) = P(Z > 0) - P(0 < Z < 2.1)$$
$$= 0.5 - 0.4821$$
$$= 0.0179$$

Case (v)

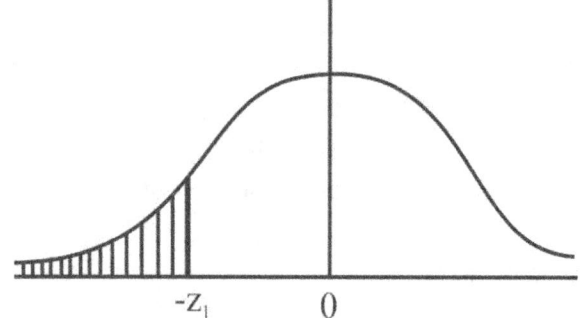

The shaded area $= P(Z \leq -Z_1)$
By symmetry

$$P(Z \leq -Z_1) = P(Z \geq Z_1)$$
$$= P(Z \geq 0) - P(0 < Z < Z_1)$$

If $Z_1 = 2.35$

$$\therefore \; P(Z \leq -2.35) = P(Z \leq 0) - P(-2.35 < Z < 0)$$
$$= P(Z \geq 0) - P(0 < Z < 2.35)$$
$$= 0.5 - 0.4906$$
$$= 0.0094$$

Case (vi)

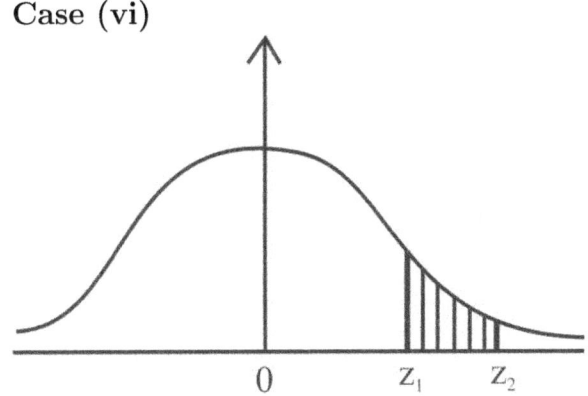

The shaded area is $P(Z_1 < Z < Z_2)$
$$= P(0 < Z < Z_2) - P(0 < Z < Z_1)$$

If $Z_1 = 1.43$ and $Z_2 = 2.22$

$$\text{The area} = P(1.43 < Z < 2.22)$$
$$= P(0 < Z < 2.22) - P(0 < Z < 1.43)$$
$$= 0.4868 - 0.4236$$
$$= 0.0632$$

Chapter 4

Case (vii)

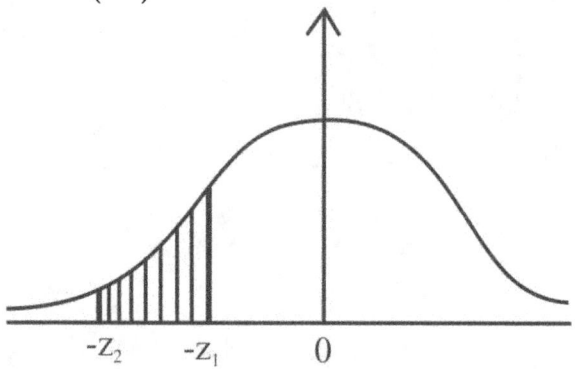

$$
\begin{aligned}
\text{The shaded area} &= P(-Z_2 < Z < -Z_1) \\
\text{By symmetry} &= P(Z_1 < Z < Z_2) \\
&= P(0 < Z < Z_2) - P(0 < Z < Z_1)
\end{aligned}
$$

If $Z_2 = 0.4$ and $Z_1 = 0.3$

$$
\begin{aligned}
\therefore P(-Z_2 < Z < -Z_1) &= P(-0.4 < Z < -0.3) \\
&= P(0.3 < Z < 0.4) \\
&= P(0 < Z < 0.4) - P(0 < Z < 0.3) \\
&= 0.1554 - 0.1179 \\
&= 0.0375
\end{aligned}
$$

Case (viii)

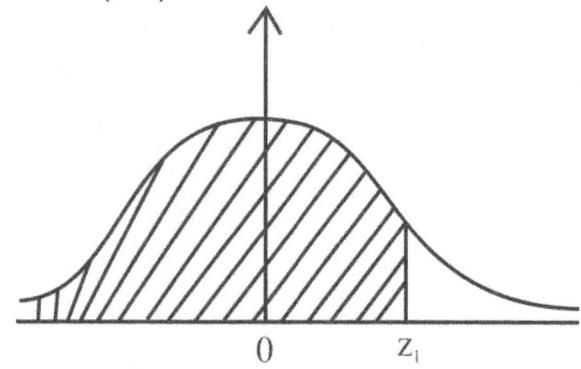

Chapter 4

The shaded area = $P(Z \leq Z_1)$
= $P(0 < Z < Z_1) + P(Z \leq 0)$
= $P(Z \leq 0) + P(0 < Z < Z_1)$

Given that $Z_1 = 0.35$

The shaded area = $P(Z \leq 0.35)$ = $P(Z \leq 0) + P(0 < Z < 0.35)$
= $0.5 + 0.1368$
= 0.6368

Case (ix)

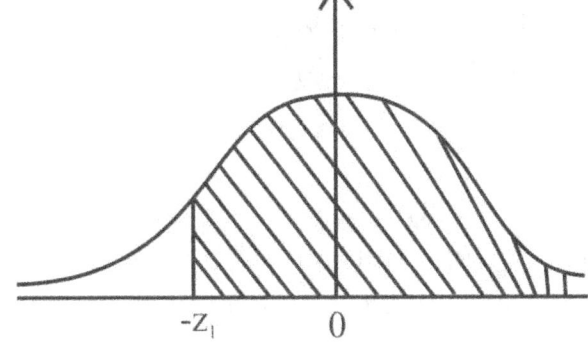

The shaded area = $P(-Z_1 < Z < 0) + P(Z \geq 0)$
= $P(0 \leq Z \leq Z_1) + P(Z \geq 0)$

If $Z_1 = 0.66$,

The shaded area = $P(-0.66 < Z \leq 0) + P(Z \geq 0)$
= $P(0 \leq Z \leq 0.66) + P(Z \geq 0)$
= $0.2454 + 0.5000$
= 0.7454

Case (x)

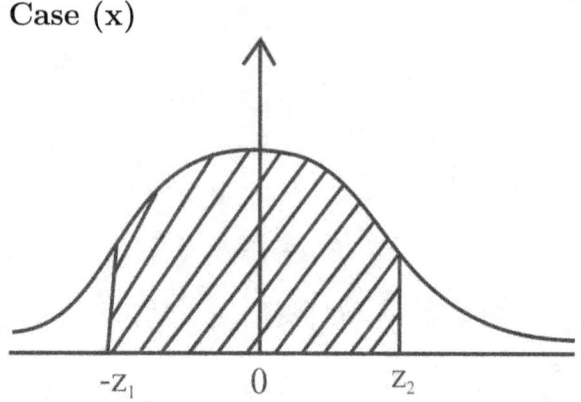

The shaded area $= P(-Z_1 < Z < Z_2)$
$= P(-Z_1 < Z < 0) + P(0 < Z < Z_2)$
$= P(0 < Z < Z_1) + P(0 < Z < Z_2)$

Given that $Z_1 = 2.51$ and $Z_2 = 3.51$

$\therefore$ The shaded area $= P(-2.51 < Z < 3.51)$
$= P(-2.51 < Z < 0) + P(0 < Z < 3.51)$
$= P(0 < Z < 2.51) + P(0 < Z < 3.51)$
$= 0.4940 + 0.4998$
$= 0.9938$

With the understanding of these ten cases, we are now in position to apply them to problems involving calculation of the area under the normal curve.

The normal distribution with mean $\mu = 0$ and $\sigma = 1$ is called the Standard Normal Distribution.

Let X has a normal distribution with mean μ and standard deviation σ, then $Z = \frac{X - \mu}{\sigma}$ is called the standardized normal distribution or the Z-score.

The Z-score is used to determine probabilities relating to random variables having normal distribution other than standard normal distribution.

Example 1:
Find y such that

(i) $P(z < y) = 0.8907$

(ii) $P(z > y) = 0.0351$

(iii) $P(|z| < y) = 0.6372$

Solution
(i)

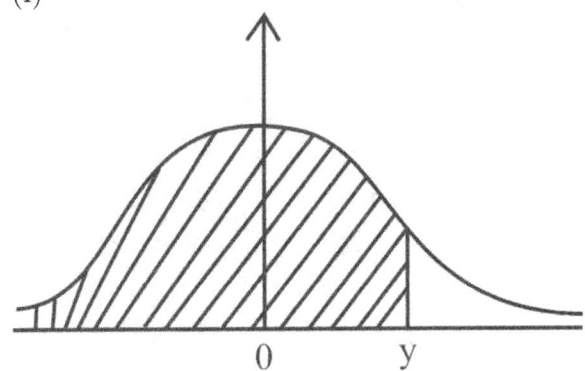

From the figure above,

$$\begin{aligned} P(z < y) &= P(0 < z < y) + P(z \leq 0) \\ &= P(0 < z < y) + 0.5 \\ P(0 < z < y) &= P(z < y) - 0.5 \\ &= 0.8907 - 0.5 \\ &= 0.3907 \end{aligned}$$

i.e. $P(0 < z < y) = 0.3907$
From the table $y = 1.23$

(ii)

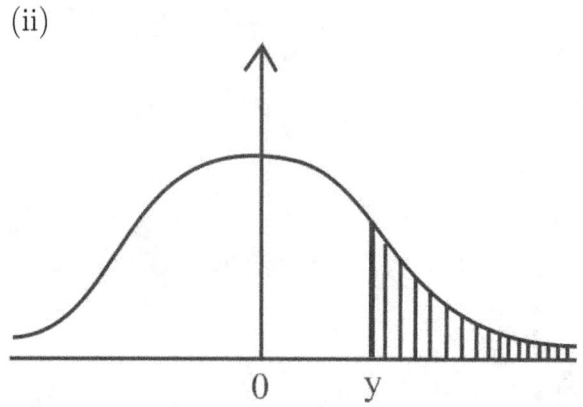

$$\begin{aligned} P(z > y) &= P(z \geq 0) - P(0 < z < y) \\ &= 0.5 - P(0 < z < y) \\ \therefore P(0 < z < y) &= 0.5 - P(z > y) \\ &= 0.5 - 0.0351 \\ &= 0.4649 \end{aligned}$$

i.e. $P(0 < z < y) = 0.4649$
From the table $y = 1.81$
(iii)

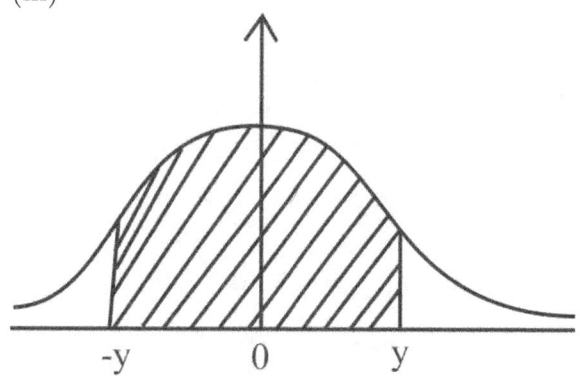

Chapter 4

$$\text{But } P(|z| < y) = P(-y < z < y)$$
$$= P(-y < z < 0) + P(0 < z < y)$$
$$= P(0 < z < y) + P(0 < z < y)$$
$$= 2P(0 < z < y)$$
$$\therefore P(0 < z < y) = \frac{1}{2}(P|z| < y)$$
$$= \frac{1}{2}(0.6372)$$
$$= 0.3186$$
$$\therefore P(0 < z < y) = 0.3186$$

From the table $y = 0.91$.

Example 2

The mean and standard deviation of a professional examination are 75 and 15 respectively. Find the scores in standard units of a candidate with (i) 65, (ii) 80, (iii) 60 marks.

Solution:
Given $\mu = 75$, $\sigma = 15$.
Let Z represent the standard score

$$\therefore Z = \frac{x - \mu}{\sigma}$$

when $x = 65$,
$$Z = \frac{65 - 75}{15} = \frac{-10}{15} = \frac{-2}{3} = -0.667$$

When $x = 80$,
$$Z = \frac{80 - 75}{15} = \frac{5}{15} = 0.333$$

when $x = 60$,
$$Z = \frac{60 - 75}{15} = \frac{-15}{15} = -1.00$$

Example 3

Use the information in example 2 to find the mark scored by a candidate with standard score of (i) 0.5, (ii) -1.5.

Solution

Given $\mu = 75$ and $\sigma = 15$.
Let Z represent the standard score

$$\therefore Z = \frac{x - \mu}{\sigma}$$

where x represents the mark of the candidate.

(i) when $Z = 0.5$, $0.5 = \dfrac{x - 75}{15}$

$$\therefore (0.5)(15) = x - 75$$

$$\therefore x = 75 + (0.5)(15)$$

$$= 75 + 7.5 = 82.5$$

(ii) when $Z = -1.5$

$$\therefore -1.5 = \frac{x - 75}{15}$$

$$\therefore -(1.5)(15) = x - 75$$

$$\therefore x = 75 - 22.5$$
$$= 52.5$$

Example 4

The marks scored by 1000 candidates in a test are normally distributed with a mean mark of 50 and a standard deviation of 10, find the proportion of candidates whose marks are

(i) below 60

Chapter 4

(ii) above 55

(iii) between 60 and 70

Solution

Given: $\mu = 50$, $\sigma = 10$, $x = 60$

$$Z = \frac{x - \mu}{\sigma} = \frac{60 - 50}{10} = 1.0$$

$$\therefore P(X < 60) = P(Z < 1.0)$$

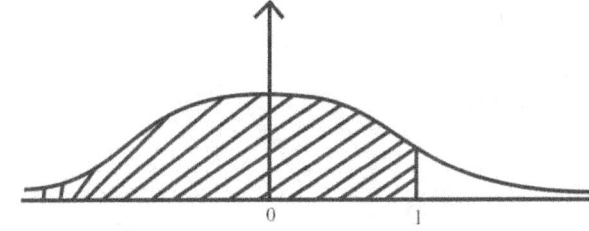

From the above figure,

$$\begin{aligned} P(Z < 1.0) &= P(Z < 0) + P(0 < Z < 1.0) \\ &= 0.5 + 0.3413 \\ &= 0.8413 \end{aligned}$$

(ii) For $x = 55$, $\sigma = 10$, $\mu = 50$

$$Z_1 = \frac{x - \mu}{\sigma} = \frac{55 - 50}{10} = \frac{5}{10} = 0.5$$

$$\therefore P(X > 55) = P(Z > 0.5)$$

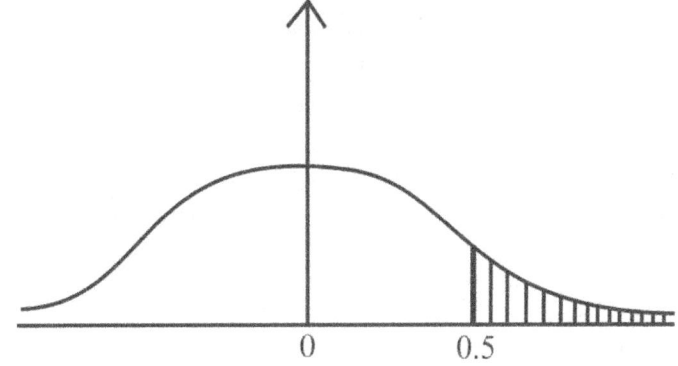

From the above figure,

$$P(Z > 0.5) = P(Z \geq 0) - P(0 < Z < 0.5)$$
$$= 0.5 - 0.1915$$
$$= 0.3085$$

(iii) When $x_1 = 60$ and $x_2 = 70$

$$Z_1 = \frac{x_1 - \mu}{\sigma} = \frac{60 - 50}{10} = 1.0$$
$$Z_2 = \frac{x_2 - \mu}{\sigma} = \frac{70 - 50}{10} = 2.0$$

$\therefore P(60 < X < 70) = P(1.0 < Z < 2.0)$

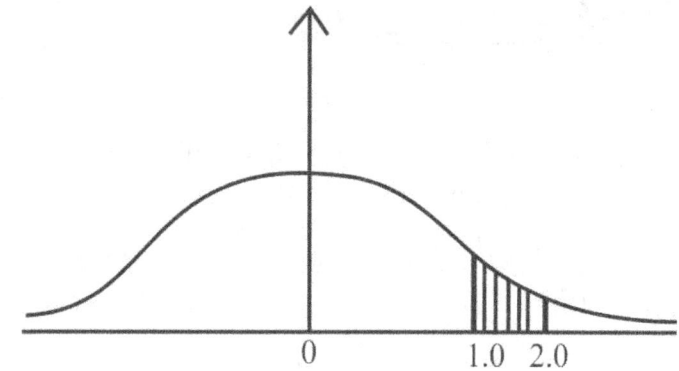

From the figure above,

$$P(60 < X < 70) = P(1 < Z < 2.0) = P(0 < Z < 2.0) - P(0 < Z < 1.0)$$
$$= 0.4772 - 0.3413$$
$$= 0.1359$$

Example 5

The marks scored in percentage by candidates in a competitive entrance examination in a particular year were normally distributed with mean 55 and standard deviation 5 respectively.

(i) If the pass mark is between 45 and 50 what percentage of the candidates passed the examination?

(ii) If 10,000 candidates sat for the examination, find the number of candidates that failed.

Solution

Given $\mu = 55$, $\sigma = 5$, $x_1 = 45$, $x_2 = 50$

$$Z_1 = \frac{x_1 - \mu}{\sigma} = \frac{45 - 55}{5} = \frac{-10}{5} = -2.0$$

$$Z_2 = \frac{x_2 - \mu}{\sigma} = \frac{50 - 55}{5} = \frac{-5}{5} = -1.0$$

$\therefore \quad P(45 < x < 50) = P(-2.0 < Z < -1.0)$

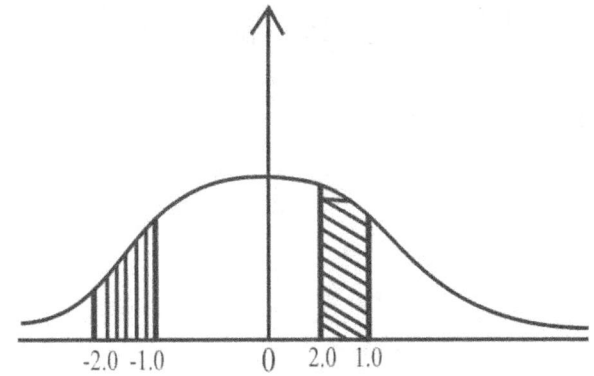

From the above figure,

$$\begin{aligned}
P(45 < x < 50) &= P(-2.0 < Z < -1.0) \\
&= P(-2.0 < Z < 0) - P(-1.0 < Z < 0) \\
&= P(0 < Z < 2) - P(0 < Z < 1) \text{ by symmetry} \\
&= 0.4772 - 0.3413 \\
&= 0.1359
\end{aligned}$$

$\therefore$ Percentage of the candidates that passed the examination
$= 0.1359 \times 100\% = 13.59\%$.

Since 10,000 sat for the examination, total number of students that passed the examination $= 10,000 \times 0.1359 = 1359$ students.

$\therefore$ No. of students that failed $= 10,000 - 1359 = 8641$.

Example 6

The life span of electric bulbs produced in a certain company is normally distributed with mean 120 hours and standard deviation 5 hours. Out of 2000 bulbs produced, how many of them would you expect to have life span ranging from 110 hours to 125 hours?

Solution

Given $\mu = 120$, $\sigma = 5$, $x_1 = 110$, $x_2 = 125$

$$Z_1 = \frac{x_1 - \mu}{\sigma} = \frac{110 - 120}{5} = \frac{-10}{5} = -2.0$$

$$Z_2 = \frac{x_2 - \mu}{\sigma} = \frac{125 - 120}{5} = \frac{5}{5} = 1.0$$

$\therefore \quad P(110 < x < 125) = P(-2.0 < Z < 1.0)$

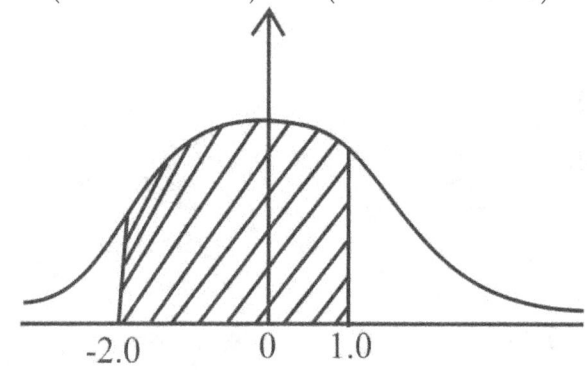

From the above figure,

$$\begin{aligned}
P(110 < x < 125) &= P(-2.0 < Z < 1.0) \\
&= P(-2.0 < Z < 0) + P(0 < Z < 1.0) \\
&= P(0 < Z < 2.0) + P(0 < Z < 1.0) \\
&= 0.4772 + 0.3413 \\
&= 0.8185
\end{aligned}$$

Since 2000 bulbs were produced, the number of bulbs with the given life span is $2000 \times 0.8185 = 1,637$.

Example 7:
If $X \sim N(3, 25)$ find $P(5 < X < 8)$.

Solution
$X \sim N(3, 25) \Rightarrow X$ is continuous random variable from a normal distribution whose mean $(\mu) = 3$ and variance $(\sigma^2) = 25$; $\sigma = \sqrt{25} = 5$.

Let $X_1 = 5$ and $X_2 = 8$.

$$\therefore Z_1 = \frac{x_1 - \mu}{\sigma} = \frac{5 - 3}{5} = \frac{2}{5} = 0.4$$

$$Z_2 = \frac{8 - 3}{5} = \frac{5}{5} = 1.0$$

$$\therefore P(X_1 < X < X_2) = P(5 < X < 8)$$
$$= P(0.4 < Z < 1.0)$$
$$= P(0 < Z < 1.0) - P(0 < Z < 0.4)$$
$$= 0.3413 - 0.1554$$
$$= 0.1859$$

Normal Approximation to Binomial distribution

The conditions under which a binomial distribution can be approximated by the normal distribution are

(i) n must be large $(n > 30)$

(ii) neither p nor q is small $(p > 0.1; q > 0.1)$; $(0 \leq p \leq 1, 0 \leq q \leq 1)$

Thus, the standard Z-score becomes

$$Z = \frac{x - \mu}{\sigma} = \frac{x - np}{\sqrt{npq}}$$

where $\mu = np$ and $\sigma = \sqrt{npq}$.

Example 1
An automobile machine produces metal bolts 40% of which are defective. If a

sample of 250 bolts is produced at a given time, find the mean and the standard deviation of the distribution.

Solution
$n = 250$, $p = \dfrac{40}{100} = \dfrac{2}{5}$, $q = 1 - p = 1 - \dfrac{2}{5} = \dfrac{3}{5}$.

$\therefore$ Mean $= \mu = np = 250 \times \dfrac{2}{5} = 100$.

$$\begin{aligned}\text{Standard deviation} &= \sqrt{npq} \\ &= \sqrt{250 \times \dfrac{2}{5} \times \dfrac{3}{5}} \\ &= \sqrt{60} = 2\sqrt{15} \\ &= 7.75\end{aligned}$$

Example 2
A fair coin is tossed 256 times. What is the probability that the number of tails occurring is above 118?

Solution
$n = 256 > 30$, probability of a tail appearing $(p) = \dfrac{1}{2}$

$$q = 1 - p = 1 - \dfrac{1}{2} = \dfrac{1}{2} > 0.1$$

Mean $(\mu) = np = 256 \times \dfrac{1}{2} = 128$

$$\begin{aligned}\text{Standard deviation } \sqrt{npq} &= \sqrt{256 \times \dfrac{1}{2} \cdot \dfrac{1}{2}} \\ &= \sqrt{64} = 8.\end{aligned}$$

Hence $\mu = 128$, $\sigma = 8$, $x = 118$

$$\therefore z_1 = \dfrac{x - \mu}{\sigma} = \dfrac{118 - 128}{8} = \dfrac{-10}{8} = -1.25$$

Chapter 4

$\therefore P(X > 118) = P(Z > -1.25)$.

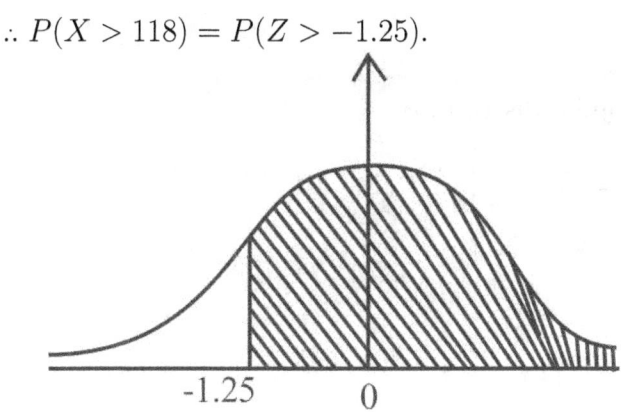

From the figure above,

$$\begin{aligned} P(X > 118) &= P(-1.25 < Z < 0) + P(Z \geq 0) \\ &= P(0 < Z < 1.25) + P(Z \geq 0) \\ &= 0.3944 + 0.5 \\ &= 0.8944 \end{aligned}$$

Negative Binomial Distribution

A random variable X has a negative binomial distribution if and only if its probability distribution is given by

$$b(x, k, \theta) = \binom{x-1}{k-1} \theta^k (1-\theta)^{x-k}$$

for $x = k, k+1, k+2, \ldots$ where X is a negative binomial random variable.

The mean of negative binomial distribution $\mu = \dfrac{k}{\theta}$ and its variance is $\sigma^2 = \dfrac{k}{\theta}\left(\dfrac{1}{\theta} - 1\right)$.

Practical Example 1:

If the probability is 0.25 that somebody will believe a rumour about the transgressions of a certain senator, find the probability that

(a) the eighth person to hear the rumour will be the fifth to believe it.

(b) the 10th person to hear the rumour will be the 9th to believe it.

Solution
(a) Here $x = 8$, $k = 5$, $\theta = 0.25$
Recall that for a negative binomial distribution

$$b(x, k, \theta) = \binom{x-1}{k-1} \theta^k (1-\theta)^{x-k}$$

$$\therefore b(8, 5, 0.25) = \binom{8-1}{5-1} (0.25)^5 (1-0.25)^{8-5}$$

$$= \binom{7}{4} (0.25)^5 (0.75)^3$$

$$= 35(0.25)^5 (0.75)^3$$

$$= 0.01442$$

(b) $x = 10$, $k = 9$, $\theta = 0.25$

$$\therefore b(10, 9, 0.25) = \binom{10-1}{9-1} (0.25)^9 (1-0.25)^{10-9}$$

$$= \binom{9}{8} (0.25)^9 (0.75)^1$$

$$= 9(0.25)^9 (0.75)$$

$$= 0.000025749$$

Example 2
If the probability of having a male child in a family is 0.4, find the probability that the family's 5th child is their third son.

Solution
$x = 5$, $k = 3$, $\theta = 0.4$

$$\therefore b(5, 3, 0.4) = \binom{5-1}{3-1} (0.4)^3 (0.6)^{5-3}$$

$$= \binom{4}{2} (0.4)^3 (0.6)^2$$

$$= 6(0.4)^3 (0.6)^2$$

$$= 0.13824$$

Geometric Distribution.

A random variable X bas a geometric distribution if and only if its probability distribution is given by $f(x,\theta) = \theta(1-\theta)^{x-1}$ for $x = 1, 2, 3, \ldots$.

Practical Example 1

The probability that a candidate will pass WASCE on any given trial is 0.5, what is the probability that a candidate will finally pass the examination on the third trial?

Solution
Here $x = 3$ and $\theta = 0.5$
By geometric distribution

$$\therefore f(x,\theta) = \theta(1-\theta)^{x-1}$$
$$\therefore f(3, 0.5) = 0.5(1-0.5)^{3-1}$$
$$= (0.5)(0.5)^2 = 0.125.$$

Hypergeometric Distribution

Definition: A random variable X bas a hypergeometric distribution, if and only if its probability distribution is given by

$$b(x; n, N, k) = \frac{\binom{k}{x}\binom{N-k}{n-x}}{\binom{N}{n}} \text{ for } x = 0, 1, 2, \ldots, n; x \leq k \text{ and } n-x \leq N-k$$

Practical Example 1

As part of an air pollution survey, an inspector decides to examine the exhaust of 6 of a company's 24 trucks. If 4 of the company's trucks emit excessive amounts of pollutants, what is the probability that none of them will be included in the inspector's sample? Give your answer to 2 decimal places.

Solution

Recall that for hypergeometric distribution,

$$b(x, n, N, k) = \frac{\binom{k}{x}\binom{N-k}{n-x}}{\binom{N}{n}}$$

Here $x = 0$, $n = 6$, $N = 24$ and $k = 4$.

$$\therefore b(0; 6, 24, 4) = \frac{\binom{4}{0}\binom{20}{6}}{\binom{24}{6}}$$

$$= 0.2880$$

$$= 0.29 \text{ to 2 decimal places}$$

Example 2

Ten applicants apply for the post of a sale manager in a company. Six of them have university degrees. If four of these applicants are randomly chosen for an interview what is the probability that two of them have college degrees? Give your answer to 4 decimal places.

Solution

Here $x = 2$, $n = 6$, $N = 10$, and $k = 4$.

Using hypergeometric distribution

$$b(2,6,10,4) = \frac{\binom{4}{2}\binom{10-4}{6-2}}{\binom{10}{6}}$$

$$= \frac{\binom{4}{2}\binom{6}{4}}{\binom{10}{6}} = \frac{6(15)}{210} = \frac{6(15)}{210} = \frac{90}{210} = \frac{3}{7}$$

$$= 0.42857$$

$$= 0.4286$$

Multinomial Distribution

The random variable $x_1, x_2, \ldots,$ and x_k have a multinomial distribution if, and only if, their joint probability distribution is given by

$$f(x_1, x_2, \ldots, x_k, n, \theta_1, \theta_2, \ldots, \theta_k) = \frac{n}{x_1! x_2! \ldots x_k!} \theta_1^{x_1} \theta_2^{x_2} \ldots \theta_k^{x_k}$$

for $x = 0, 1, \ldots, n$ for each i where $\sum_{i=1}^{k} x_i = n$ and $\sum_{i=1}^{k} \theta_k = 1$.

Practical Example 1

In a particular town on a Sunday night, channel 4 has 25 percent of the viewing audience, channel 3 has 20 percent of the viewing audience and channel 1 has 50 percent of the viewing audience. Find the probability that among ten television viewers in that town randomly chosen on Sunday night, four will be watching channel 4, three will be watching channel 3 and one will be watching channel 1.

Solution

By multinomial distribution,

$$
\begin{aligned}
f(x_1, x_2, \ldots, x_k, n, \theta_1, \theta_2, \ldots, \theta_k) &= \frac{n!}{x_1! x_2! \ldots x_k!} \theta_1^{x_1} \theta_2^{x_2} \ldots \theta_k^{x_k} \\
f(4, 3, 1, 10, 0.25, 0.20, 0.50) &= \frac{10!}{4! 3! 1!} (0.25)^4 \cdot (0.20)^3 (0.50)^1 \\
&= 25200 (0.25)^4 (0.20)^3 (0.50)^1 \\
&= 0.39375
\end{aligned}
$$

Chapter 5

Moment of a Random Variable (Discrete and Continuous)

Definition 1
The rth moment of a discrete random variable X about the mean is defined as
$$\mu_r = E[(X-\mu)^r] = \sum_{i=1}^{n}(x_i-\mu)^r f(x^r), r=0,1,2\ldots$$

Definition 2
The rth moment of a discrete random variable X about the origin is defined as
$$\mu'_r = E(X^r) = \sum_{i=1}^{n} x_i^r f(x_i),\ r=0,1,2,\ldots$$

When $r=1$, $\mu'_1 = E(X^1) = E(X) = \mu$

$\therefore\ \mu'_1 = \mu\ \Rightarrow$ the first moment about the origin is the mean

Relationship between μ_r and μ'_r.
In calculating rth moment about the mean (μ_r), the following relationships have been found useful.

Chapter 5

Relationship Between Moments About the Mean and the Moments About the Origin

1. 1st Moment about the mean is given by

$$\mu_1 = E(X - \mu) = E(X) - E(\mu) = \mu'_1 - \mu'_1 = 0$$

2. 2nd Moment about the mean is given by

$$\begin{aligned}
\mu_2 = E(X - \mu)^2 &= E(X^2 - 2\mu X + \mu^2) \\
&= E(X^2) - 2\mu E(X) + E(\mu^2) \\
&= \mu'_2 - 2\mu \cdot \mu + \mu^2 \\
&= \mu'_2 - 2\mu^2 + \mu^2 \\
&= \mu'_2 - \mu^2 \\
&= \mu'_2 - (\mu'_1)^2 \qquad \mu = \mu'_1)
\end{aligned}$$

3. 3rd Moment about the mean is given by

$$\begin{aligned}
\mu_3 = E(X - \mu)^3 &= E[X^3 - 3X^2\mu + 3X\mu^2 - \mu^3] \\
&= E(X^3) - 3E(X^2)\mu + 3E(X)\mu^2 - E(\mu^3) \\
&= \mu'_3 - 3\mu'_2\mu + 3\mu \cdot \mu^2 - \mu^3 \\
&= \mu'_3 - 3\mu'_2\mu + 2\mu^3 \\
&= \mu'_3 - 3\mu'_2\mu'_1 + 2(\mu'_1)^3
\end{aligned}$$

4. Using Binomial expansion, the 4th Moment about the mean is given by

$$\begin{aligned}
\mu_4 = E(X - \mu)^4 &= E(X^4 - 4\mu X^3 + 6\mu^2 X^2 - 4\mu^3 X + \mu^4) \\
&= E(X^4) - 4\mu E(X^3) + 6\mu^2 E(X^2) - 4\mu^3 E(X) + E(\mu^4) \\
&= \mu'_4 - 4\mu\mu'_3 + 6\mu^2\mu'_2 - 4\mu^3\mu + \mu^4 \\
&= \mu'_4 - 4\mu\mu'_3 + 6\mu^2\mu'_2 - 4\mu^4 + \mu^4 \\
&= \mu'_4 - 4\mu\mu'_3 + 6\mu^2\mu'_2 - 3\mu^4 \\
&= \mu'_4 - 4\mu'_1\mu'_3 + 6\mu'_2(\mu'_1)^2 - 3(\mu'_1)^4
\end{aligned}$$

Chapter 5

5. 5th moment about the mean is given by

$$\begin{aligned}\mu_5 &= E(X-\mu)^5 \\ &= E(X^5) - 5\mu X^4 + 10\mu^2 X^3 - 10\mu^3 X^2 + 5\mu^4 X - \mu^5) \\ &= E(X^5) - 5\mu E(X^4) + 10\mu^2 E(X^3) - 10\mu^3 E(X^2) + 5\mu^4 E(X) - E(\mu^5) \\ &= \mu_5' - 5\mu\mu_4' + 10\mu^2\mu_3' - 10\mu^3\mu_2' + 5\mu^4(\mu) - \mu^5 \\ &= \mu_5' - 5\mu\mu_4' + 10\mu^2\mu_3' - 10\mu^3\mu_2' + 4\mu^5 \\ &= \mu_5' - 5\mu_1'\mu_4' + 10(\mu_1')^2\mu_3' - 10(\mu_1')^3\mu_2' + 4(\mu_1')^5\end{aligned}$$

6. The 6th moment about the mean is given by

$$\begin{aligned}\mu_6 &= E(X-\mu)^6 \\ &= E(X^6) - 6X^5\mu + 15X^4\mu^2 - 20X^3\mu^3 + 15X^2\mu^4 - 6X\mu^5 + \mu^6) \\ &= E(X^6) - 6E(X^5)\mu + 15E(X^4)\mu^2 - 20E(X^3)\mu^3 \\ &\quad + 15E(X^2)\mu^4 - 6E(X)\mu^5 + \mu^6 \\ &= \mu_6' - 6\mu_5'\mu + 15\mu_4'\mu^2 - 20\mu_3'\mu^3 + 15\mu_2'\mu^4 - 6\mu\cdot\mu^5 + \mu^6 \\ &= \mu_6' - 6\mu_5'\mu + 15\mu_4'\mu^2 - 20\mu_3'\mu^3 + 15\mu_2'\mu^4 - 5\mu^6 \\ &= \mu_6' - 6\mu_5'\mu_1' + 15\mu_4'(\mu_1')^2 - 20\mu_3'(\mu_1')^3 + 15\mu_2'(\mu_1')^4 - 5(\mu_1')^6\end{aligned}$$

Note: (1) Other higher moments about the mean can be found in terms of moments about the origin using binomial expansion as shown above.
(2) The first moment about the origin given by μ_1' is the same as μ i.e., $\mu = \mu_1'$.
In summary,

(1) $\mu_1 = 0$

(2) $\mu_2 = \mu_2' - (\mu_1')^2$

(3) $\mu_3 = \mu_3' - 3\mu_2'\mu_1' + 2(\mu_1')^3$

(4) $\mu_4 = \mu_4' - 4\mu_1'\mu_3' + 6\mu_2'(\mu_1')^2 - 3(\mu_1')^4$

(5) $\mu_5 = \mu_5' - 5\mu_1'\mu_4' + 10(\mu_1')^2\mu_3' - 10\mu_2'(\mu_1')^3 + 4(\mu_1')^5$

(6) $\mu_6 = \mu_6' - 6\mu_5'\mu_1' + 15\mu_4'(\mu_1')^2 - 20\mu_3'(\mu_1')^3 + 15\mu_2'(\mu_1')^4 - 5(\mu_1')^6$.

Chapter 5

where μ_r is the moment about the mean
and μ'_r is the moment about the origin; $r = 1, 2, 3, \ldots$

Example 1
Find the first three moments about the origin of the random variable X with the probability distribution below and use your result to find the corresponding moments about the mean.

x	-4	2	6
$f(x)$	1/4	1/2	1/4

Solution
The first moment about the origin $\mu_1 = E(X) = \mu$.
From the above distribution

$$\begin{aligned}
\mu'_1 = E(X) &= \sum x f(x) \\
&= \left(-4 \times \frac{1}{4}\right) + \left(2 \times \frac{1}{2}\right) + \left(6 \times \frac{1}{4}\right) \\
&= -1 + 1 + \frac{3}{2} \\
&= 1.5
\end{aligned}$$

The second moment about the origin $\mu_2 = E(X^2)$

$$\begin{aligned}
\mu'_2 = E(X^2) &= \sum x^2 f(x) = -4^2 \left(\frac{1}{4}\right) + 2^2 \left(\frac{1}{2}\right) + 6^2 \left(\frac{1}{4}\right) \\
&= \left(16 \times \frac{1}{4}\right) + \left(4 \times \frac{1}{2}\right) + \left(36 \times \frac{1}{4}\right) \\
&= 4 + 2 + 9 \\
&= 15
\end{aligned}$$

The third moment about the origin is

$$\mu'_3 = E(X^3) = \sum x^3 f(x)$$
$$= \left((-4)^3 \times \frac{1}{4}\right) + \left(2^3 \times \frac{1}{2}\right) + \left(6^3 \times \frac{1}{4}\right)$$
$$= \left(-64 \times \frac{1}{4}\right) + \left(8 \times \frac{1}{2}\right) + \left(216 \times \frac{1}{4}\right)$$
$$= -16 + 4 + 54$$
$$= 42$$

Now,

$$\text{First moment about the mean } \mu_1 = E(X - \mu) = E(X) - \mu = 0$$
$$\text{The second moment about the mean } \mu_2 = \mu'_2 - (\mu'_1)^2$$
$$= 15 - (1.5)^2$$
$$= 15 - 2.25$$
$$= 12.75$$
$$\text{The third moment about the mean } \mu_3 = \mu'_3 - 3\mu'_2\mu'_1 + 2(\mu'_1)^3$$
$$= 42 - \left(3 \times 15 \times \frac{3}{2}\right) + 2\left(\frac{3}{2}\right)^3$$
$$= 42 - \frac{135}{2} + \frac{27}{4}$$
$$= -18.75$$

Example 2

Given that X is a random variable with probability density function $f(x) = \dfrac{x}{3}$ for $x = 0, 1, 2$. Find the first four moments about the origin. Use your result to find the second and third moments about the mean.

Solution

Chapter 5

First moment about the origin is $E(X) = \mu'_1$,

$$\therefore \mu'_1 = E(X) = \sum xf(x) = \left(0 \times \frac{0}{3}\right) + \left(1 \times \frac{1}{3}\right) + \left(2 \times \frac{2}{3}\right)$$

$$= \frac{1}{3} + \frac{4}{3} = \frac{5}{3} = 1\frac{2}{3}$$

Second moment about the origin is $E(X^2) = \mu'_2$,

$$\therefore \mu'_2 = E(X^2) = \sum x^2 f(x) = 0 + \left(1 \times \frac{1}{3}\right) + \left(2^2 \times \frac{2}{3}\right)$$

$$= \frac{1}{3} + \frac{8}{3}$$

$$= \frac{9}{3} = 3$$

Third moment about the origin is $E(X^3) = \mu'_3$

$$\therefore \mu'_3 = E(X^3) = \sum x^3 f(x) = 0 + \left(1 \times \frac{1}{3}\right) + \left(2^3 \times \frac{2}{3}\right)$$

$$= \frac{1}{3} + \frac{16}{3}$$

$$= \frac{17}{3} = 5\frac{2}{3}$$

Fourth moment about the origin is $E(X^4) = \mu'_4$

$$\therefore \mu'_4 = E(X^4) = \sum x^4 f(x) = 0 + \left(1 \times \frac{1}{3}\right) + \left(2^4 \times \frac{2}{3}\right)$$

$$= \frac{1}{3} + \frac{32}{3}$$

$$= \frac{33}{3} = 11$$

Second moment about the mean

$$\mu_2 = \mu'_2 - (\mu'_1)^2$$

$$= 3 - \left(\frac{5}{3}\right)^2 = 3 - \frac{25}{9} = \frac{27 - 25}{9} = \frac{2}{9}$$

Chapter 5

Third moment about the mean

$$\begin{aligned}\mu_3 &= \mu_3' - 3\mu_2'\mu_1' + 2(\mu_1')^3 \\ &= \frac{17}{3} - \left(3 \times 3 \times \frac{5}{3}\right) + 2\left(\frac{5}{3}\right)^3 = \frac{17}{3} - 15 + \frac{250}{27} \\ &= \frac{153 - 405 + 250}{27} = \frac{403 - 405}{27} = \frac{-2}{27}\end{aligned}$$

Moment of a Continuous Random Variable
Definitions:

(1) The rth moment of a continuous random variable X about the mean is given by

$$\mu_r = E[(x-\mu)^r] = \int_{-\infty}^{\infty} (x-\mu)^r f(x) \quad r = 0, 1, 2, \ldots$$

Hence the 1st, 2nd, 3rd,... and rth moment about the mean is $\mu_1, \mu_2, \mu_3, \ldots, \mu_r$.

(2) The rth moment of a continuous random variable X about the origin is given by

$$\mu_r' = E(X^r) = \int_{-\infty}^{\infty} x^r f(x) dx; \quad r = 0, 1, 2, \ldots$$

Note: The relationship between μ_r and μ_r' for discrete r.v. is the same for continuous r.v.

Example 1
A continuous random variable X has a probability density function

$$f(x) = \begin{cases} \dfrac{2x-1}{6} & , 0 < x < 2 \\ 0 & , \text{otherwise} \end{cases}$$

Find the first three moments about the origin.
Solution

The first moment about the origin

$$\begin{aligned}
\mu'_1 = E(X) &= \int_0^2 xf(x)dx \\
&= \int_0^2 \frac{x(2x+1)}{6}dx \\
&= \frac{1}{6}\int_0^2 (2x^2 + x)dx \\
&= \frac{1}{6}\left[\frac{2}{3}x^3 + \frac{1}{2}x^2\right]_0^2 \\
&= \frac{1}{6}\left[\frac{2}{3}(2)^3 + \frac{1}{2}(2)^2 - 0\right] \\
&= \frac{1}{6}\left(\frac{16}{3} + 2\right) \\
&= \frac{22}{18} = \frac{11}{9} = 1.22
\end{aligned}$$

$\therefore$ The first moment about the origin $\mu'_1 = 1.22$.
The second moment about the origin is

$$\begin{aligned}
\mu'_2 = E(X^2) &= \int_0^2 x^2 f(x) \\
&= \int_0^2 \frac{x^2(2x+1)}{6} \\
&= \frac{1}{6}\int_0^2 (2x^3 + x^2)dx \\
&= \frac{1}{6}\left[\frac{1}{2}x^4 + \frac{1}{3}x^3\right]_0^2 \\
&= \frac{1}{6}\left[\left[\frac{1}{2}(2^4) + \frac{1}{3}(2^3)\right] - 0\right]
\end{aligned}$$

$$= \frac{1}{6}\left[\frac{16}{2} + \frac{8}{3}\right]$$

$$= \frac{1}{6}\left(\frac{64}{6}\right)$$

$$= \frac{64}{36}$$

$$= \frac{16}{9} = 1.78$$

∴ The second moment about the origin $\mu_2' = 1.78$.
The third moment about the origin is

$$\mu_3' = E(X^3) = \frac{1}{6}\int_0^2 x^3(2x+1)dx$$

$$= \frac{1}{6}\int_0^2 (2x^4 + x^3)dx$$

$$= \frac{1}{6}\left[\frac{2}{5}x^5 + \frac{1}{4}x^4\right]_0^2$$

$$= \frac{1}{6}\left\{\left[\frac{2}{5}(32) + \frac{1}{4}(16)\right] - 0\right\}$$

$$= \frac{1}{6}\left\{\left[\frac{64}{5} + 4\right]\right\} = \frac{1}{6}\left(\frac{84}{5}\right)$$

$$= \frac{84}{30} = 2.8$$

$$= 2.8$$

∴ The third moment about the origin $\mu_3' = 2.8$.

Example 2
The continuous random variable X has a probability density function

$$f(x) = \begin{cases} \frac{1}{4} & , 0 < x < 4 \\ 0 & , \text{otherwise} \end{cases}$$

Find the first three moments about the origin. Use your result to find the second and third moment about the mean.

Solution

The first moment about the origin is

$$\mu'_1 = E(X) = \int_0^4 x f(x) dx$$

$$= \int_0^4 \frac{x}{4} dx = \frac{x^2}{8}\bigg]_0^4 = \frac{16}{8} = 2.$$

$\therefore \quad \mu'_1 = \mu = 2.$

The second moment about the origin is

$$\mu'_2 = E(X^2) = \int_0^4 x^2 f(x) dx$$

$$= \int_0^4 \frac{x^2}{4} dx$$

$$= \frac{x^3}{12}\bigg|_0^4 = \frac{64}{12} = \frac{16}{3} = 5.33$$

$\therefore \quad \mu'_2 = 5.33.$

The third moment about the origin is

$$\mu'_3 = E(X^3) = \int_0^4 x^3 f(x) dx$$

$$= \int_0^4 \frac{x^3}{4} dx$$

$$= \frac{x^4}{16}\bigg|_0^4 = \frac{256}{16} = 16.$$

$\therefore \quad \mu'_3 = 16.$

Chapter 5

The second moment about the mean is

$$\mu_2 = \mu_2^1 - (\mu_1')^2 = \frac{16}{3} - 2^2$$
$$= \frac{16}{3} - 4$$
$$= \frac{4}{3}$$
$$= 1\frac{1}{3}$$

$\therefore \mu_2 = 1\frac{1}{3}.$

The third moment about the mean is

$$\mu_3 = \mu_3^1 - 3\mu_2^1\mu_1' + 2(\mu_1')^3$$
$$= 16 - \left(3 \times \frac{16}{3} \times 2\right) + (2 \times 2^3)$$
$$= 16 - 32 + 16 = 0$$

Factorial Moments

The rth factorial moment about the origin is defined as:

$$\mu_{[r]}' = E[X(X-1)(X-2)\ldots(X-r+1)] \quad \text{for } r = 1, 2, 3, \ldots$$
$$= \sum x(x-1)(x-2)\ldots(x-r+1)f(x)$$

(for discrete random variables)

For continuous random variables

$$\mu_{[r]}' = E[X(X-1)(X-2)\ldots(X-r+1)]$$
$$= \int_x x(x-1)(x-2)\ldots(x-r+1)f(x)dx$$

Relationship Between Factorial Moments and Moments about the Origin.

1st factorial moment about the origin $\mu_{[1]}' = E(X) = \mu_1^1$

Chapter 5

2nd factorial moment about the origin
$\mu'_{[2]} = E(X(X-1)] = E(X^2) - E(X) = \mu'_2 - \mu'_1$

3rd factorial moment about the origin
$$\begin{aligned}\mu'[3] &= E[(X(X-1)(X-2))] \\ &= E(X^3 - 3X^2 + 2X) \\ &= E(X^3) - 3E(X^2) + 2E(X) \\ &= \mu'_3 - 3\mu'_2 + 2\mu'_1\end{aligned}$$

4th factorial moment about the origin
$$\begin{aligned}\mu'_{[4]} &= E[X(X-1)(X-2)(X-3)] \\ &= E(X^4 - 6X^3 + 11X^2 - 6X) \\ &= E(X^4) - 6E(X^3) + 11E(X^2) - 6E(X) \\ &= \mu'_4 - 6\mu'_3 + 11\mu'_2 - 6\mu'_1\end{aligned}$$

5th factorial moment about the origin
$$\begin{aligned}\mu'_{[5]} &= E[(X)(X-1)(X-2)(X-3)(X-4)] \\ &= E(X^5 - 10X^4 + 13X^3 + 38X^2 + 24X) \\ &= E(X^5) - 10E(X^4) + 13E(X^3) + 38E(X^2) + 24E(X) \\ &= \mu'_5 - 10\mu'_4 + 13\mu'_3 + 38\mu'_2 + 24\mu'_1\end{aligned}$$

From the above, factorial moment can be expressed as algebraic sum of moments about the origin.

$$\mu'_{[r]} = E[(X(X-1)\cdots(X-r+1)) \quad r = 1,.2,3,\ldots$$

1st factorial moment about the origin $= \mu_{[1]} = \mu'_1$
2nd factorial moment about the origin $= \mu_{[2]} = \mu'_2 - \mu'_1$
3rd factorial moment about the origin $= \mu_{[3]} = \mu'_3 - 3\mu'_2 + 2\mu'_1$
4th factorial moment about the origin $= \mu_{[4]} = \mu'_4 - 6\mu'_3 + 11\mu'_2 - 6\mu'_1$
5th factorial moment about the origin $= \mu_{[5]} = \mu'_5 - 10\mu'_4 + 13\mu'_3 + 38\mu'_2 + 24\mu'_1$

Chapter 5

Example 1
A random variable X has the following discrete probability density function:

$$X = \begin{cases} 0 & \text{with probability } \dfrac{1}{8} \\ 1 & \text{with probability } \dfrac{1}{4} \\ 2 & \text{with probability } \dfrac{5}{8} \end{cases}$$

Use the information above to find:

(i) the first three moments about the origin

(ii) the second and third factorial moments about the origin.

Solution
Writing the discrete probability density function given in tabular form we obtain the table below:

x	0	1	2
$f(x)$	$\dfrac{1}{8}$	$\dfrac{1}{4}$	$\dfrac{5}{8}$

$$\begin{aligned} \text{Ist moment about the origin } \mu_1' &= E(X) \\ &= \sum x f(x) \\ &= \left(0 \times \frac{1}{8}\right) + \left(1 \times \frac{1}{4}\right) + \left(2 \times \frac{5}{8}\right) \\ &= \frac{1}{4} + \frac{5}{4} \\ &= \frac{3}{2} \end{aligned}$$

$$
\begin{aligned}
\text{2nd moment about the origin } \mu'_2 &= E(X^2) \\
&= \sum x^2 f(x) \\
&= \left(0 \times \frac{1}{8}\right) + \left(1 \times \frac{1}{4}\right) + \left(2^2 \times \frac{5}{8}\right) \\
&= \frac{1}{4} + \frac{5}{2} = \frac{11}{4} \\
\text{3rd moment about the origin } \mu'_3 &= E(X^3) \\
&= \sum x^3 f(x) \\
&= \left(0 \times \frac{1}{8}\right) + \left(1 \times \frac{1}{4}\right) + \left(2^3 \times \frac{5}{8}\right) \\
&= \frac{1}{4} + 5 \\
&= \frac{21}{4}
\end{aligned}
$$

$$
\begin{aligned}
\text{2nd factorial moment about the origin} &= \mu'_{[2]} = \mu'_2 - \mu'_1 \\
&= \frac{11}{4} - \frac{3}{2} \\
&= \frac{5}{4} = 1\frac{1}{4} \\
\text{3rd factorial moment about the origin} &= \mu'_{[3]} = \mu'_3 - 3\mu'_2 + 2\mu'_1 \\
&= \frac{21}{4} - \left(3 \times \frac{11}{4}\right) + \left(2 \times \frac{3}{2}\right) \\
&= \frac{21}{4} - \frac{33}{4} + 3 \\
&= \frac{33}{4} - \frac{33}{4} \\
&= 0
\end{aligned}
$$

Example 2
If X is a random variable whose probability density function is given by

$$f(x) = \begin{cases} 1 &, 0 \leq x \leq 1 \\ 0 &, \text{elsewhere} \end{cases}$$

find the first, second, third and fourth factorial moment about the origin

Solution

Since factorial moment is the algebraic sum of moments about the origin, we shall find all the moments about the origin.

$\mu'_r = (r = 1, 2, 3, 4)$

$\mu'_r = E(X^r) = \int x^r f(x) dx.$

$$\begin{aligned}
\text{1st moment about the origin } \mu'_1 = E(X) &= \int_0^1 x f(x) dx \\
&= \int_0^1 x(1) dx \\
&= \left.\frac{1}{2}x^2\right|_0^1 = \frac{1}{2}(1-0) = \frac{1}{2} \\
\text{2nd moment about the origin } \mu'_2 = E(X^2) &= \int_0^1 x^2 f(x) dx \\
&= \int_0^1 x^2(1) dx \\
&= \int_0^1 x^2 dx \\
&= \left.\frac{1}{3}x^3\right|_0^1 = \frac{1}{3}(1-0) = \frac{1}{3}
\end{aligned}$$

3rd moment about the origin $\mu'_3 = E(X^3) = \int_0^1 x^3(1)dx$

$$= \frac{x^4}{4}x^2 \Big|_0^1 = \frac{1}{4}$$

4th moment about the origin $\mu'_4 = E(X^4) = \int_0^1 x^4 dx$

$$= \frac{1}{5}x^5 \Big|_0^1 = \frac{1}{5}$$

Hence $\mu'_1 = \frac{1}{2}$, $\mu'_2 = \frac{1}{3}$, $\mu'_3 = \frac{1}{4}$, $\mu'_4 = \frac{1}{5}$.

$\therefore$ 1st factorial moment about the origin $\mu'_{[1]} = \mu'_1 = \frac{1}{2}$

2nd factorial moment about the origin $\mu'_{[2]} = \mu'_2 - \mu'_1 = \frac{1}{3} - \frac{1}{2} = -\frac{1}{6}$

3rd factorial moment about the origin $\mu'_{[3]} = \mu'_3 - 3\mu'_2 + 2\mu'_1$

$$= \frac{1}{4} - 3(\frac{1}{3}) + 2(\frac{1}{2})$$

$$= \frac{1}{4}$$

4th factorial moment $\mu'_{[4]} = \mu'_4 - 6\mu'_3 + 11\mu'_2 - 6\mu'_1$

$$= \frac{1}{5} - 6(\frac{1}{4}) + 11(\frac{1}{3}) - 6(\frac{1}{2})$$

$$= \frac{1}{5} - \frac{3}{2} + \frac{11}{3} - 3$$

$$= \frac{-19}{30}$$

Chapter 6

Continuous Random Variable

Definition:
A random variable X is said to be continuous if its choice of values falls within an interval i.e. if $x \in [a, b]$ for $x_1, x_2, \ldots, x_n \in X$ where $[a, b] = a \le x \le b$.

It can also be defined as that random variable which takes on a non-countably infinite number of values.

Probability Density Function of a Continuous Random Variable

Properties:
If X is a continuous random variable, a function $f(x)$ is said to be a continuous probability density function if it satisfies the following conditions

(i) $f(x) \ge 0$

(ii) $\int_{-\infty}^{\infty} f(x) dx = 1.$

Note: (i) If X is a continuous random variable, we shall assume unless otherwise stated that

$$P(a \le X \le b) = P(a \le X < b) = P(a < X \le b) = P(a < X < b) = \int_a^b f(x) dx$$

(ii) Probability density function is abbreviated p.d.f.

Example 1
A continuous random variable X has a probability density function

$$f(x) = \begin{cases} x+t & ; \ 0 < x < 3 \\ 0 & ; \ \text{elsewhere} \end{cases}$$

(a) Find (i) the value of t, (ii) $P(1 < X < 3)$, (iii) $P(0 < X < 5)$.

Solution
For $f(x)$ to be a probability density function

$$\int_a^b f(x)dx = 1.$$

$$\begin{aligned}
\int_0^3 f(x)dx &= \int_0^3 (x+t)dx \\
&= \left| \frac{1}{2}x^2 + tx \right|_0^3 \\
&= \left. \frac{1}{2}x^2 \right|_0^3 + \left. tx \right|_0^3 \\
&= \frac{1}{2}(3^2 - 0) + t(3-0) \\
&= \frac{9}{2} + 3t.
\end{aligned}$$

Since $f(x)$ is a p.d.f.

$$\therefore \int_0^3 f(x)dx = 1$$

Hence, $\dfrac{9}{2} + 3t = 1$

$$3t = 1 - \dfrac{9}{2}$$
$$= -\dfrac{7}{2}$$
$$\therefore\ t = -\dfrac{7}{6}$$

$$f(x) = \begin{cases} x - \dfrac{7}{6} & ;\ 0 < x < 3 \\ 0 & ;\ \text{elsewhere} \end{cases}$$

(ii) $\begin{aligned}[t] P(1 < X < 3) &= \int_1^3 f(x)\,dx \\ &= \int_1^3 \left(x - \dfrac{7}{6}\right) dx \\ &= \left|\dfrac{1}{2}x^2 - \dfrac{7}{6}x\right|_1^3 \\ &= \dfrac{1}{2}x^2 \Big|_1^3 - \dfrac{7}{6}x \Big|_1^3 \\ &= \dfrac{1}{2}(3^2 - 1) - \dfrac{7}{6}(3 - 1) \\ &= \dfrac{1}{2}(8) - \dfrac{7}{6}(2) \\ &= 4 - \dfrac{7}{3} = \dfrac{5}{3} \end{aligned}$

(iii) $P(0 < X < 5) = P(0 < X < 3)$; $(4 < x < 5)$ is outside the given range

$$= \int_0^3 f(x)dx = \int_0^3 \left(x - \frac{7}{6}\right) dx \Big|_0^3$$

$$= \frac{1}{2}x^2 \Big|_0^3 - \frac{7}{6}x \Big|_0^3$$

$$= \frac{1}{2}(3^2 - 0) - \frac{7}{6}(3 - 0)$$

$$= \frac{9}{2} - \frac{7}{2} = 1$$

$$= 1$$

Example 2

Given that X is a random variable with probability density function

$$f(x) = \begin{cases} kx^2 & ; \ 1 < X < 3 \\ 0 & ; \ \text{elsewhere} \end{cases}$$

Find (i) the constant k, (ii) $P(1 < X < 2)$, (iii) $P(-1 < X < -3)$

Solution

(i) Since $f(x)$ is a p.d.f.,

$$\int_1^3 kx^2 = 1$$

$$\frac{1}{3}kx^3 \Big|_1^3 = 1$$

i.e. $\frac{1}{3}k(3^3 - 1) = 1$

$$\frac{26}{3}k = 1$$

Chapter 6

$$\therefore k = \frac{3}{26}$$

$$f(x) = \begin{cases} \dfrac{3}{26}x^2 & ; \ 1 < X < 3 \\ 0 & ; \ \text{elsewhere} \end{cases}$$

(ii) $P(1 < X < 2) = \displaystyle\int_1^2 \dfrac{3}{26}x^2 dx$

$$= \dfrac{1}{26}x^3 \Big|_1^2 = \dfrac{1}{26}(2^3 - 1^3)$$

$$= \dfrac{1}{26}(8-1) = \dfrac{7}{26}$$

(iii) $P(-1 < X < -3) = 0$ $\because$ $-1 < X < -3$ is completely outside the given range.

Mean and Variance of a Continuous Random Variable

Definition: For a continuous random variable X with probability density function $f(x)$, the expectation of X is defined as

$$E(X) = \int_{-\infty}^{\infty} x f(x) dx$$

where $E(X) = \mu$.

$$Var(X) = E[(X - \mu)]^2 = E(X^2) - \mu^2$$

where

$$E(X^2) = \int_{-\infty}^{\infty} x^2 f(x) dx.$$

Example 1

Given that

$$f(x) = \begin{cases} x & ; \ 0 \leq x \leq 1 \\ 2 - x & ; \ 1 \leq x \leq 2 \\ 0 & ; \ \text{elsewhere} \end{cases}$$

Chapter 6

Is $f(x)$ a p.d.f.? Find the mean of X.

Solution:
For $f(x)$ to be a p.d.f.,
$$\int_a^b f(x)dx = 1.$$

Here,
$$\begin{aligned}
\int_X f(x)dx &= \int_0^1 x\,dx + \int_1^2 (2-x)dx = \frac{x^2}{2}\Big|_0^1 + 2x\Big|_1^2 - \frac{1}{2}x^2\Big|_1^2 \\
&= \frac{1}{2}(1^2 - 0) + 2(2-1) - \frac{1}{2}(2^2 - 1^2) \\
&= \frac{1}{2}(1) + 2 - \frac{1}{2}(3) \\
&= \frac{1}{2} + 2 - \frac{3}{2} \\
&= 1
\end{aligned}$$

Since $\int f(x)dx = 1$; $f(x)$ is a p.d.f.

The mean
$$\begin{aligned}
E(X) = \int xf(x)dx &= \int_0^1 x(x)dx + \int_1^2 x(2-x)dx \\
&= \int_0^1 x^2 dx + \int_1^2 (2x - x^2)dx \\
&= \frac{1}{3}x^3\Big|_0^1 + x^2\Big|_1^2 - \frac{1}{3}x^3\Big|_1^2 \\
&= \frac{1}{3}(1^3 - 0) + (2^2 - 1^2) - \frac{1}{3}(2^3 - 1^3) \\
&= \frac{1}{3}(1) + (4-1) - \frac{1}{3}(8-1) \\
&= \frac{1}{3} + 3 - \frac{7}{3} \\
&= 1
\end{aligned}$$

Example 2
A random variable X has a p.d.f. of
$$f(x) = \begin{cases} k & ; \quad 0 < X < 4 \\ 0 & ; \quad \text{elsewhere} \end{cases}$$

Find (i) the constant k, (ii) mean and variance of X.
Since $f(x)$ is a p.d.f.

$$\therefore \int_0^4 k\,dx = 1$$

$$\therefore kx\big|_1^4 = 1$$

$$k(4-0) = 1 \Rightarrow 4k = 1 \therefore k = \frac{1}{4}$$

$$\therefore f(x) = \begin{cases} \frac{1}{4} & ; 0 < x < 4 \\ 0 & ; \text{elsewhere} \end{cases}$$

The mean
$$\begin{aligned}
E(X) = \int_0^4 xf(x)dx = \int_0^4 x\left(\frac{1}{4}\right)dx &= \int_0^4 \frac{1}{4}x\,dx \\
&= \frac{1}{8}x^2\bigg|_0^4 \\
&= \frac{1}{8}(4^2 - 0) \\
&= \frac{16}{8} \\
&= 2
\end{aligned}$$

$$\therefore E(X) = \mu = 2.$$

$$E(X^2) = \int_0^4 x^2 f(x)dx = \int_0^4 \frac{1}{4}x^2 dx = \frac{1}{12}x^3 \Big|_0^4 = \frac{1}{12}(4^3 - 0)$$

$$= \frac{64}{12} = \frac{16}{3}$$

$$\therefore Var(X) = E(X^2) - \mu^2$$

$$= \frac{16}{3} - 2^2 = \frac{16}{3} - 4$$

$$= \frac{4}{3}$$

Example 3

The density function of a continuous random variable X is given by

$$f(x) = \begin{cases} \dfrac{1}{\ln 3^{x^2}} & ; \text{for } 1 < X < 3 \\ 0 & ; \text{otherwise} \end{cases}$$

find (i) $Var(X)$ (ii) $E(x^2 + 2x - 1)$.

Solution

$$E(X) = \int_x x f(x) dx$$

$$= \int_1^3 x \left(\frac{1}{\ln 3^{x^2}}\right) dx$$

$$= \int_1^3 x \left[\frac{1}{x^2 \ln 3}\right] dx$$

$$= \frac{1}{\ln 3} \int_1^3 \frac{x}{x^2} dx = \frac{1}{\ln 3} \int \frac{1}{x} dx$$

$$= \frac{1}{\ln 3}[\ln x |_1^3]$$

$$= \frac{\ln 3 - \ln(1)}{\ln 3} = \frac{\ln 3}{\ln 3}$$

$$\therefore E(X) = 1$$

Chapter 6

$$\begin{aligned}
E(X^2) &= \int_x x^2 f(x)dx \\
&= \int_1^3 x^2 \left(\frac{1}{\ln 3^{x^2}}\right) dx. \\
&= \int_1^3 x^2 \left[\frac{1}{x^2 \ln 3}\right] = \int_1^3 \frac{1}{\ln 3} dx = \left(\frac{1}{\ln 3}\right) \int_1^3 dx \\
&= \frac{1}{\ln 3} x \Big|_1^3 \\
&= \frac{1}{\ln 3}[3-1] = \frac{2}{\ln 3} \\
\therefore Var(X) &= E(X^2) - [E(X)]^2 \\
&= \frac{2}{\ln 3} - 1^2 \\
&= 1.820 - 1 = 1.820 \\
\therefore Var(X) &= 1.820 \\
E(X^2 + 2X - 1) &= \int_X (x^2 + 2x - 1)f(x)dx. \\
&= \int_1^3 (x^2 + 2x - 1)\left(\frac{1}{\ln 3^{x^2}}\right) dx \\
&= \int_1^3 \frac{x^2}{\ln 3^{x^2}} dx + 2\int_1^3 x\left(\frac{1}{\ln 3^{x^2}}\right) dx - \int_1^3 \frac{1}{\ln 3^{x^2}} dx. \\
&= \int_1^3 \frac{x^2}{x^2 \ln 3} dx + 2\int_1^3 \frac{x}{x^2 \ln 3} dx - \int_1^3 \frac{1}{x^2 \ln 3} dx \\
&= \frac{1}{\ln 3}\int_1^3 dx + 2\int_1^3 \frac{1}{x \ln 3} dx - \frac{1}{\ln 3}\int_1^3 (x^{-2})dx \\
&= \frac{1}{\ln 3}(x|_1^3) + \frac{2}{\ln 3}(\ln x|_1^3) + \frac{1}{\ln 3}(\frac{1}{x}|_1^3) \\
&= \frac{1}{\ln 3}(3-1) + \frac{2}{\ln 3}(\ln 3 - \ln 1) + \frac{1}{\ln 3}\left(\frac{1}{3} - 1\right) \\
&= \frac{2}{\ln 3} + 2 - \frac{2}{3\ln 3} = 2 + \frac{4}{3\ln 3} = 3.214 \text{ approx}
\end{aligned}$$

Chapter 6

Example 4
Given that
$$f(x) = \begin{cases} \dfrac{4}{\pi(1+x^2)} & ; 0 < x < 1 \\ 0 & ; \text{otherwise} \end{cases}$$
is a probability density function of a random variable X, show that
$$E(X) = \frac{\ln 4}{\pi}$$

Solution

$$\begin{aligned}
E(X) = \int_X xf(x)dx &= \int_0^1 x\left(\frac{4}{\pi(1+x^2)}\right)dx \\
&= \frac{4}{\pi}\int_0^1 \frac{x}{1+x^2}dx. \\
&= \frac{4}{\pi}\left[\frac{1}{2}\ln(1+x^2)\right] \\
&= \frac{2}{\pi}\ln(1+x^2)\bigg|_0^1 \\
&= \frac{2}{\pi}(\ln 2 - \ln 1) \\
&= \frac{2}{\pi}(\ln 2 - 0) \\
&= \frac{2\ln 2}{\pi} = \frac{\ln 2^2}{\pi} \\
&= \frac{\ln 4}{\pi} \quad \text{proved.}
\end{aligned}$$

Probability Distribution Function
Definition 1
If $f(x)$ and $F(x)$ are the values of the probability density function and probability distribution function of X respectively at x, then $P(a \leq X \leq b) = F(b) - F(a)$ and for any real constants a and b with $a \leq b$, $f(x) = \dfrac{dF(x)}{dx}$

where the derivative exists.

Definition 2
If X is a continuous random variable, the function given by

$$F(x) = P(X \leq x) = \int_{-\infty}^{x} f(t)dt \text{ for } -\infty < x < \infty$$
$$= P(-\infty < X \leq x)$$

where $f(t)$ is the value of the probability density function of X at t.

Example 1
Given that X is a continuous random variable with probability density function

$$f(x) = \begin{cases} 3e^{-3x} & ; x \geq 0 \\ 0 & ; \text{elsewhere} \end{cases}$$

Find (i) its probability distribution function. (ii) Use your result to find $P(0.5 \leq X \leq 1)$.

Solution
For $x > 0$,

$$F(x) = \int_{-\infty}^{x} f(t)dt$$
$$= \int_{0}^{x} 3e^{-3t}dt = -e^{-3t}\Big|_{0}^{x}$$
$$= -\left[e^{-3x} - e^{-0}\right]$$
$$\therefore F(x) = 1 - e^{-3x}$$

But
$$P(a \leq X \leq b) = F(b) - F(a)$$
$$\therefore P(0.5 \leq X \leq 1) = F(1) - F(0.5)$$

Since $F(X) = 1 - e^{-3x}$; $F(1) = 1 - e^{-3}$,

$$\begin{aligned}
F(0.5) = 1 - e^{-3(0.5)} &= 1 - e^{-1.5} \\
\therefore \quad F(1) - F(0.5) &= (1 - e^{-3}) - (1 - e^{-1.5}) \\
&= e^{-1.5} - e^{-3} \\
&= 0.173
\end{aligned}$$

Example 2
The distribution function of a random variable X is given by

$$F(x) = \begin{cases} 0 & ; x < 0 \\ \dfrac{x^4}{64} & ; 0 \leq x \leq 4 \\ 1 & ; x \geq 4 \end{cases}$$

Obtain (i) the probability density function of X,
(ii) $P(-2 < X < 1)$.

Solution
By definition,

$$f(x) = \frac{dF(x)}{dx} = \frac{d}{dx} \begin{cases} 0 & ; x < 0 \\ \dfrac{x^4}{64} & ; 0 \leq x \leq 4 \\ 1 & ; x \geq 4 \end{cases}$$

Chapter 6

$$= \begin{cases} 0 & ; x < 0 \\ \dfrac{4x^3}{64} & ; 0 \le x \le 4 \\ 0 & ; x \ge 4 \end{cases}$$

$$\therefore \ f(x) = \begin{cases} \dfrac{x^3}{16} & ; 0 \le x < 4 \\ 0 & ; \text{elsewhere} \end{cases}$$

$$\begin{aligned}
\therefore \ P(-2 < X < 1) &= P(0 < X < 1) \\
&= \int_0^1 f(x)\,dx \\
&= \int_0^1 \dfrac{x^3}{16}\,dx \\
&= \left.\dfrac{x^4}{64}\right|_0^1 = \dfrac{1}{64}(1-0) = \dfrac{1}{64}
\end{aligned}$$

The Mode of a Distribution

For a discrete distribution, the mode is the value of X for which the probability is maximum. For a continuous distribution, the mode can be obtained by method of calculus.

Example 1

The probability mass function of a random variable X is given by

$$f(x) = \begin{cases} \frac{2}{7} & ; x = 0 \\ \frac{1}{7} & ; x = 1 \\ \frac{3}{7} & ; x = 2 \\ \frac{1}{7} & ; x = 3. \end{cases}$$

Find the mode of the distribution.

Solution:
Since the mode is the value of X for which the probability is maximum, the value of X for which the probability is maximum in the above distribution is 2. Hence the mode is 2.

Example 2
The probability mass function of a random variable Y is given by

$$f(x) = \begin{cases} \frac{1}{3} & , x = 1 \\ \frac{1}{6} & , x = 2 \\ \frac{1}{6} & , x = 3 \\ \frac{1}{3} & , x = 4. \end{cases}$$

Show that the above distribution is bi-modal.

Solution
The probability is maximum at $x = 1$ and $x = 4$.

Chapter 6

∴ The modes of the distribution are 1 and 4.
The modes are two in number, hence the distribution is bi-modal.

Example 3
Given a random variable X with probability density function

$$f(x) = \begin{cases} \dfrac{1}{2}xe^{-x/2} &, x \geq 0 \\ 0 &, \text{elsewhere} \end{cases}$$

Find the mode of the distribution.

Solution
Since $f(x) = \dfrac{1}{2}xe^{-x/2}$

$$\begin{aligned} f'(x) &= -\dfrac{1}{4}xe^{-x/2} + \dfrac{1}{2}e^{-x/2} \\ &= \dfrac{1}{2}e^{-x/2}(-\dfrac{1}{2}x + 1) \end{aligned} \quad (1)$$

When $f'(x) = 0$,

$$\dfrac{1}{2}e^{-x/2}(1 - \dfrac{1}{2}x) = 0 \quad (2)$$

$$\Rightarrow \quad e^{-x/2} = 0 \text{ or } 1 - \dfrac{1}{2}x = 0$$

When $e^{-x/2} = 0$, $x = \infty$ which is undefined
when $1 - \dfrac{1}{2}x = 0$, $x = 2$.
From (1)

$$\begin{aligned} f''(x) &= -\dfrac{1}{4}e^{-x/2}(1 - \dfrac{1}{2}x) - \dfrac{1}{2}(\dfrac{1}{2}e^{-x/2}) \\ &= -\dfrac{1}{4}e^{-x/2}\left(1 - \dfrac{1}{2}x\right) - \dfrac{1}{4}e^{-x/2} \end{aligned}$$

when $x = 2$,

$$\begin{aligned} f''(x) &= -\frac{1}{4}[e^{-1}(1-1)] - \frac{1}{4}e^{-1} \\ &= 0 - \frac{1}{4e} = -\frac{1}{4e} < 0 \end{aligned}$$

Since $f''(x) < 0$, $\exists$ a maximum point.
$\therefore$ $x = 2$ is a maximum point.
Hence the mode of the distribution is 2.

The Median and Quartiles of a Distribution
Definition: The median of a continuous random with probability density function $f(x)$ is given by

$$\int_{-\infty}^{m} f(x)dx \leq \frac{1}{2} \quad \text{or} \quad \int_{m}^{\infty} f(x)dx \geq \frac{1}{2}$$

where m is the median.

Example.
Find the (i) median, (ii) the first quartile, (iii) second quartile and (iv) semi interquartile range of a continuous random variable with probability density function

$$f(x) = \begin{cases} 2e^{-2x} &, x \geq 0 \\ 0 &, \text{elsewhere} \end{cases}$$

Solution

(i) By definition,

$$\text{Median} = \int_0^m f(x)dx = 0.5$$

i.e. $$\int_0^m 2e^{-2x}dx = 0.5$$

i.e. $$-e^{-2x}\Big|_0^m = 0.5$$

$$\therefore -(e^{-2m} - e^0) = 0.5$$

$$-e^{-2m} + 1 = 0.5$$

$$-e^{-2m} = -0.5$$

$$\therefore e^{-2m} = 0.5$$

Take log of both sides of the equation

$$\therefore \log_e e^{-2m} = \log_e 0.5 = \log_e \frac{1}{2}$$

$$\therefore -2m = \log \frac{1}{2} = \log 2^{-1}$$

$$\therefore -2m = -\log 2$$

$$\therefore 2m = \log 2$$

$$m = \frac{\log_e 2}{2} = \frac{0.6931}{2} = 0.346.$$

∴ The median of the distribution is 0.346.

(ii) The 1st quartile of a continuous random variable with probability distribution $f(x)$ is given by

$$\int_{-\infty}^{q_1} f(x)dx \le 0.25.$$

or

$$\int_{q_1}^{\infty} f(x)dx \ge 0.25.$$

Solution:

The first quartile $= \int_0^{q_1} 2e^{-2/x} dx = 0.25$

i.e. $\int_0^{q_1} 2e^{-2x} dx = \dfrac{1}{4}$

$\therefore \quad -e^{-2x}\Big|_0^{q_1} = \dfrac{1}{4}$

$-(e^{-2q_1} - e^0) = \dfrac{1}{4}$

$\therefore \quad -e^{-2q_1} + 1 = \dfrac{1}{4}$

i.e. $e^{-2q_1} = \dfrac{3}{4}$

$\begin{aligned}
\therefore \quad \log_e e^{-2q_1} &= \log_e \dfrac{3}{4} \\
\therefore \quad -2q_1 &= \log_e 0.75 \\
\therefore \quad q_1 &= -\dfrac{1}{2} \log_e 0.75 \\
\therefore \quad q_1 &= 0.144
\end{aligned}$

Hence the 1st quartile (Q_1) is 0.144.

(iii) The third quartile of a random variable X with probability density function $f(x)$ satisfies the inequality

$$\int_{-\infty}^{q_3} f(x)dx \leq 0.75$$

or

$$\int_{q_3}^{\infty} f(x)dx \geq 0.75$$

Solution:

The third quartile $= \int_0^{q_3} 2e^{-2x} dx = 0.75$

$$\text{i.e.} \int_0^{q_3} 2e^{-2x} dx = \frac{3}{4}$$

$$\text{i.e.} \left. -e^{-2x} \right|_0^{q_3} = \frac{3}{4}$$

$$\therefore -\left(e^{-2q_3} - e^0\right) = \frac{3}{4}$$

$$\therefore e^{-2q_3} + 1 = \frac{3}{4}$$

$$\therefore -e^{-2q_3} = -\frac{1}{4}$$

$$\therefore e^{-2q_3} = \frac{1}{4}$$

$$\text{i.e.} \log_e e^{-2q_3} = \log_e \frac{1}{4}$$

$$\therefore -2q_3 = \log_e 2^{-2}$$

$$\Rightarrow -2q_3 = -2\log_e 2$$

$$\Rightarrow q_3 = \log_e 2 = 0.6931$$

$\therefore$ The third quartile is 0.6931.

(iv) By definition, semi interquartile range is

$$\frac{Q_3 - Q_1}{2} = \frac{0.6931 - 0.1439}{2}$$
$$= 0.2746.$$

Mean and Variance of Some Continuous Probability Distributions

(1) **Uniform Distribution or Rectangular Distribution**
   Definition: A continuous random variable X is said to be uniformly distributed or has rectangular distribution if its probability density function

Chapter 6

is given by
$$f(x) = \begin{cases} \dfrac{1}{b-a} & ; a < x < b \\ 0 & ; \text{otherwise} \end{cases}$$

The mean of the distribution is given by

$$\begin{aligned}
E(X) = \int_a^b x f(x) dx &= \int_a^b x \left(\frac{1}{b-a}\right) dx \\
&= \frac{1}{b-a} \int_a^b x \, dx \\
&= \frac{1}{b-a} \left(\frac{x^2}{2}\bigg|_a^b\right) \\
&= \frac{1}{b-a} \left[\frac{b^2 - a^2}{2}\right] \\
&= \frac{1}{b-a} \left[\frac{(b-a)(b+a)}{2}\right] \\
&= \frac{b+a}{2}.
\end{aligned}$$

$\therefore$ The mean of the distribution $= E(X) = \dfrac{b+a}{2}$.

$$\begin{aligned}
E(X^2) = \int_a^b x^2 f(x) dx &= \int_a^b x^2 \left(\frac{1}{b-a}\right) dx \\
&= \frac{1}{b-a} \int_a^b x^2 \, dx \\
&= \frac{1}{b-a} \left[\frac{x^3}{3}\bigg|_a^b\right] \\
&= \frac{1}{b-a} \left[\frac{b^3 - a^3}{3}\right]
\end{aligned}$$

Chapter 6

$$\begin{align}
&= \frac{1}{3(b-a)}[(b-a)(a^2+ab+b^2)]\\
&= \frac{a^2+ab+b^2}{3}\\
Var(X) &= E(X^2) - [E(X)]^2\\
&= \frac{a^2+ab+b^2}{3} - \left(\frac{b+a}{2}\right)^2\\
&= \frac{a^2+ab+b^2}{3} - \frac{(b^2+2ab+a^2)}{4}\\
&= \frac{4(a^2+ab+b^2) - 3(b^2+2ab+a^2)}{12}\\
&= \frac{4a^2+4ab+4b^2 - 3b^2 - 6ab - 3a^2}{12}\\
&= \frac{b^2-2ab+a^2}{12} = \frac{(b-a)^2}{12}
\end{align}$$

(2) **Exponential Distribution**

A continuous random variable X is said to have an exponential distribution if its probability density function is given by

$$f(x) = \begin{cases} \lambda e^{-\lambda x} & ; x \geq 0 \\ 0 & ; \text{otherwise} \end{cases}$$

The mean of the distribution is $\dfrac{1}{\lambda}$

Proof

$$\begin{align}
E(X) = \int_0^\infty xf(x)dx &= \int_0^\infty x(\lambda e^{-\lambda x}dx \quad (**)\\
&= \lambda \int_0^\infty xe^{-\lambda x}dx
\end{align}$$

Put $y = \lambda x$, $x = \dfrac{y}{\lambda}$, $dx = \dfrac{dy}{\lambda}$

$$\begin{aligned}
\therefore \quad E(X) &= \lambda \int_0^\infty \frac{y}{\lambda} e^{-y} \frac{dy}{\lambda} \\
&= \frac{1}{\lambda} \int_0^\infty y e^{-y} dy \\
&= \frac{1}{\lambda} \int_0^\infty y^{2-1} e^{-y} dy \\
&= \frac{1}{\lambda} \Gamma_{(2)} \\
&= \frac{1}{\lambda}
\end{aligned}$$

The mean of exponential distribution $= \dfrac{1}{\lambda}$

$$\begin{aligned}
E(X^2) &= \int_0^\infty x^2 (\lambda e^{-\lambda x}) dx \\
&= \int_0^\infty \left(\frac{y}{\lambda}\right)^2 (\lambda e^{-y}) \frac{dy}{\lambda} \quad \text{(using } (**)\text{)} \\
&= \frac{1}{\lambda^2} \int_0^\infty y^2 e^{-y} dy \\
&= \frac{1}{\lambda^2} \int_0^\infty y^{3-1} e^{-y} dy \\
&= \frac{1}{\lambda^2} \Gamma_{(3)} \\
&= \frac{2}{\lambda^2}
\end{aligned}$$

$$\begin{aligned}
\therefore \quad Var(X) &= E(X^2) - [E(X)]^2 \\
&= \frac{2}{\lambda^2} - \frac{1}{\lambda^2} \\
&= \frac{1}{\lambda^2}
\end{aligned}$$

Hence the variance of the distribution is $\dfrac{1}{\lambda^2}$.

(3) Normal Distribution

A continuous random variable X is said to have a normal distribution if its probability density function is

$$f(x) = \dfrac{1}{\sigma\sqrt{2\pi}} e^{-\frac{1}{2}\left(\frac{x-u}{\alpha}\right)^2}, \quad -\infty < x < \infty \tag{1}$$

$$= \dfrac{1}{\sigma\sqrt{2\pi}} e^{-\frac{1}{2\sigma^2}(x-u)^2}$$

The mean

$$E(X) = \int_{-\infty}^{\infty} x f(x) dx$$

$$= \int_{-\infty}^{\infty} x \dfrac{1}{\sigma\sqrt{2\pi}} e^{-\frac{1}{2}\{\frac{x-\mu}{\sigma}\}^2} dx$$

Put $t = \dfrac{x-\mu}{\sigma}$

$\Rightarrow \quad x = \mu + \sigma t$ and $dx = \sigma dt$.

$$E(X) = \dfrac{1}{\sigma\sqrt{2\pi}} \int_{-\infty}^{\infty} (\mu + \sigma t) e^{-\frac{1}{2}t^2} \sigma dt$$

$$= \dfrac{1}{\sqrt{2\pi}} \int_{-\infty}^{\infty} (\mu + \sigma t) e^{-\frac{1}{2}t^2} dt$$

$$= \dfrac{1}{\sqrt{2\pi}} \int_{-\infty}^{\infty} \mu e^{-\frac{1}{2}t^2} dt + \dfrac{\sigma}{\sqrt{2\pi}} \int_{-\infty}^{\infty} t e^{-\frac{1}{2}t^2} dt$$

$$= \dfrac{\mu}{\sqrt{2\pi}} \int_{-\infty}^{\infty} e^{-\frac{1}{2}t^2} dt + \dfrac{\sigma}{\sqrt{2\pi}} \int_{-\infty}^{\infty} t e^{-\frac{1}{2}t^2} dt$$

But $\dfrac{1}{\sqrt{2\pi}} e^{-\frac{1}{2}t^2}$ is a probability density function of a standard normal distribution.

$$\therefore \quad E(X) = \mu.1 + \dfrac{\sigma}{\sqrt{2\pi}} \int_{-\infty}^{\infty} t e^{-\frac{1}{2}t^2} dt$$

$$E(X) = \mu + \left(\dfrac{\sigma}{\sqrt{2\pi}} \times 0\right) = \mu$$

Hence the mean of a normal distribution $= E(X) = \mu$

Using the same transformation, $t = \dfrac{x - \mu}{\sigma} \Rightarrow x = \mu + \sigma t$ and $dx = \sigma dt$.

$$\begin{aligned}
\text{The variance} &= E(X - \mu)^2 \\
&= \frac{1}{\sigma\sqrt{2\pi}} \int_{-\infty}^{\infty} (x - \mu)^2 e^{-\frac{1}{2}\left(\frac{x-\mu}{\sigma}\right)^2} dx \\
&= \frac{1}{\sigma\sqrt{2\pi}} \int_{-\infty}^{\infty} (\sigma t)^2 e^{-\frac{1}{2}t^2} \sigma dt \\
&= \frac{\sigma^3}{\sigma\sqrt{2\pi}} \int_{-\infty}^{\infty} t^2 e^{-\frac{1}{2}t^2} dt \\
&= \frac{\sigma^2}{\sqrt{2\pi}} \left[-t e^{-\frac{1}{2}t^2} + \int_{-\infty}^{\infty} e^{-\frac{1}{2}t^2} dt \right] \quad \text{(integration by parts)} \\
&= \frac{\sigma^2}{\sqrt{2\pi}} \left(-t e^{-\frac{1}{2}t^2}\right)_{-\infty}^{\infty} + \frac{\sigma^2}{\sqrt{2\pi}} \int_{-\infty}^{\infty} e^{-\frac{1}{2}t^2} dt \\
&= 0 + \sigma^2 \left(\frac{1}{\sqrt{2\pi}} \int_{-\infty}^{\infty} e^{-\frac{1}{2}t^2} dt \right) \\
&= \sigma^2 \times 1 = \sigma^2
\end{aligned}$$

Hence the variance of a random variable having normal distribution is σ^2.

(4) Gamma Distribution

A continuous random variable X is said to have Gamma distribution if its probability density function is given by

$$f(x) = \begin{cases} \dfrac{x^{\alpha-1} e^{-\frac{1}{\beta}x}}{\beta^{\alpha} \Gamma_{\alpha}} & ; \; x > 0; (\alpha, \beta > 0) \\ 0 & ; \; \text{elsewhere.} \end{cases}$$

Chapter 6

The mean of the distribution

$$\begin{aligned}
E(X) &= \int_0^\infty xf(x)dx \\
&= \int_0^\infty x\left[\frac{x^{\alpha-1}e^{\frac{-1}{\beta}x}}{\beta^\alpha \Gamma_\alpha}\right]dx \\
&= \frac{1}{\beta^\alpha \Gamma_\alpha}\int_0^\infty x(x^{\alpha-1})e^{\frac{-1}{\beta}x}dx \\
&= \frac{1}{\beta^\alpha \Gamma_\alpha}\int_0^\infty x^\alpha e^{\frac{-1}{\beta}x}dx
\end{aligned}$$

Put $y = \frac{1}{\beta}x \Rightarrow x = \beta y$. $\therefore dx = \beta dy$

$$\begin{aligned}
E(X) &= \frac{1}{\beta^\alpha \Gamma_\alpha}\int_0^\infty (\beta y)^\alpha e^{-y}\beta dy \\
&= \frac{1}{\beta^\alpha \Gamma_\alpha}\beta^\alpha \beta \int_0^\infty y^\alpha e^{-y}dy
\end{aligned}$$

Recall that
$$\Gamma_\alpha = \int_0^\infty x^{\alpha-1}e^{-x}dx = (\alpha-1)\Gamma_{\alpha-1}$$

$$\therefore E(X) = \frac{1}{\beta^\alpha \Gamma_\alpha}\beta^\alpha \beta \int_0^\infty y^\alpha e^{-y}dy = \frac{\beta \Gamma_{\alpha+1}}{\Gamma_\alpha} = \frac{\beta \alpha \Gamma_\alpha}{\Gamma_\alpha} = \alpha\beta$$

Hence the mean of Gamma distribution is $\alpha\beta$.

$$\begin{aligned}
E(X^2) &= \int_0^\infty x^2 f(x)dx \\
&= \int \frac{x^2(x^{\alpha-1}e^{-\frac{1}{\beta}x})}{\beta^\alpha \Gamma_\alpha}dx \\
&= \frac{1}{\beta^\alpha \Gamma_\alpha}\int_0^\infty (\beta y)^2(\beta y)^{\alpha-1}e^{-y}\beta dy
\end{aligned}$$

Chapter 6

$$\begin{aligned}
&= \frac{1}{\beta^\alpha \Gamma_\alpha} \int_0^\infty \beta^2 \beta^\alpha y^{\alpha+1} e^{-y} dy \\
&= \frac{\beta^2 \beta^\alpha}{\beta^\alpha \Gamma_\alpha} \Gamma_{\alpha+2} \\
&= \frac{\beta^2 \beta^\alpha (\alpha+1)\Gamma_{\alpha+1}}{\beta^\alpha \Gamma_\alpha} \\
&= \frac{\beta^2 (\alpha+1)\alpha \Gamma_\alpha}{\Gamma_\alpha} \\
&= \alpha(\alpha+1)\beta^2 \\
Var(X) &= E(X^2) - [E(X)]^2 \\
&= \alpha(\alpha+1)\beta^2 - (\alpha\beta)^2 \\
&= \alpha^2\beta^2 + \alpha\beta^2 - \alpha^2\beta^2 \\
&= \alpha\beta^2
\end{aligned}$$

Hence the variance of Gamma distribution is $\alpha\beta^2$.

(5) **Beta Distribution**

A continuous random variable X is said to have Beta distribution if its probability density function is given by

$$f(x) = \begin{cases} \dfrac{x^{a-1}(1-x)^{b-1}}{B(a,b)} &, \ a > 0; \ 0 < x < 1 \\ 0 &, \ \text{elsewhere.} \end{cases}$$

where

$$\begin{aligned}
B(a,b) &= \int_0^1 x^{a-1}(1-x)^{b-1} dx \\
&= \frac{\Gamma_{(a)} \Gamma_{(b)}}{\Gamma_{a+b}}
\end{aligned}$$

Consider

$$E(X^k) = \frac{1}{B(a,b)} \int_0^1 x^k x^{a-1}(1-x)^{b-1} dx$$

$$= \frac{1}{B(a,b)} \int_0^1 x^{k+a-1}(1-x)^{b-1} dx$$

$$= \frac{1}{B(a,b)} B(k+a, b) = \frac{\Gamma_{k+a}\Gamma_b}{\Gamma_{(k+a+b)}} \times \frac{1}{B(a,b)}$$

$$= \frac{\Gamma_{(k+a)}\Gamma_b}{\Gamma_{(k+a+b)}} \times \frac{\Gamma_{a+b}}{\Gamma_a \times \Gamma_b} = \frac{\Gamma_{(k+a)} \times \Gamma_{(a+b)}}{\Gamma_{(k+a+b)} \times \Gamma_a}$$

$$= \frac{\Gamma_{(k+a)}\Gamma_{(a+b)}}{\Gamma_a \Gamma_{(k+a+b)}}$$

When $k = 1$,

$$E(X) = \frac{\Gamma_{(a+1)}\Gamma_{(a+b)}}{\Gamma_a \Gamma_{(a+b+1)}} = \frac{a\Gamma_a \Gamma_{(a+b)}}{\Gamma_a(a+b)\Gamma_{(a+b)}} = \frac{a\Gamma_a\Gamma_{(a+b)}}{(a+b)\Gamma_a\Gamma_{(a+b)}} = \frac{a}{a+b}$$

The mean of the distribution $= \dfrac{a}{a+b}$.

When $k = 2$

$$E(X^2) = \frac{\Gamma_{(a+2)} \cdot \Gamma_{(a+b)}}{\Gamma_a \cdot \Gamma_{a+b+2}}$$

$$= \frac{\Gamma_{(a+2)}\Gamma_{(a+b)}}{(a+b)(a+b+1)\Gamma_a \cdot \Gamma_{a+b}}$$

$$= \frac{a(a+1)\Gamma_a \cdot \Gamma_{a+b}}{(a+b)(a+b+1)\Gamma_a \cdot \Gamma_{a+b}} = \frac{a(a+1)}{(a+b)(a+b+1)}$$

$$\therefore Var(X) = E(X^2) - [E(X)]^2$$

$$= \frac{a(a+1)}{(a+b)(a+b+1)} - \left[\frac{a}{a+b}\right]^2$$

$$= \frac{ab}{(a+b)^2(a+b+1)} \quad \text{(the variance of Beta Distribution)}$$

(6) **Weibull Distribution** A random variable X has a Weibull distribution if and only if its probability density function is given by:

$$f(x) = \begin{cases} \alpha\beta x^{\beta-1} e^{-\alpha x^\beta} & , x > 0 \\ 0 & , \text{elsewhere} \end{cases}$$

Find the mean and variance of a random variable with Weibull distribution.

Let X represent a r.v. with Weibull distribution.

$$\therefore \text{ Mean } (\mu) = E(X) = \int x f(x) dx$$

$$= \int_0^\infty x \alpha\beta x^{\beta-1} e^{-\alpha x^\beta} dx$$

$$= \alpha\beta \int_0^\infty x^\beta e^{-\alpha x^\beta} dx$$

Put $t = \alpha x^\beta$, $\therefore x = \left(\dfrac{t}{\alpha}\right)^{\frac{1}{\beta}} = \dfrac{1}{\alpha^{1/\beta}} t^{1/\beta}$.

$\therefore dx = \dfrac{1}{\beta \alpha^{1/\beta}} t^{\frac{1}{\beta}-1} dt.$

Using the above transformation,

$$E(X) = \alpha\beta \int_0^\infty \left[\left(\dfrac{t}{\alpha}\right)^{\frac{1}{\beta}}\right]^\beta e^{-t} \dfrac{1}{\beta \alpha^{1/\beta}} t^{\frac{1}{\beta}-1} dt$$

$$= \alpha\beta \int_0^\infty \dfrac{t}{\alpha} \cdot \dfrac{1}{\alpha^{1/\beta}\beta} (t^{\frac{1}{\beta}-1}) e^{-t} dt$$

$$= \dfrac{\alpha\beta}{\alpha \cdot \alpha^{1/\beta}\beta} \int_0^\infty t^{1/\beta} e^{-t} dt = \dfrac{1}{\alpha^{1/\beta}} \Gamma\left(1 + \dfrac{1}{\beta}\right)$$

$$= \alpha^{-\frac{1}{\beta}} \Gamma\left(1 + \dfrac{1}{\beta}\right)$$

Hence,

$$\mu = E(X) = \alpha^{-1/\beta} \Gamma\left(1 + \dfrac{1}{\beta}\right)$$

Chapter 6

Now,

$$
\begin{aligned}
E(X^2) &= \int x^2 f(x)\,dx \\
&= \int_0^\infty x^2 \alpha\beta x^{\beta-1} e^{-\alpha x^\beta}\,dx \\
&= \alpha\beta \int_0^\infty x^2 x^{\beta-1} e^{-\alpha x^\beta}\,dx. \\
&= \alpha\beta \int_0^\infty x^{\beta+1} e^{-\alpha x^\beta}\,dx \\
&= \alpha\beta \int_0^\infty \left[\left(\frac{t}{\alpha}\right)^{\frac{1}{\beta}}\right]^{\beta+1} e^{-t} \frac{1}{\beta \alpha^{1/\beta}} (t^{\frac{1}{\beta}-1}) e^{-t}\,dt \\
&= \alpha\beta \int_0^\infty \frac{1}{\alpha^{1+\frac{1}{\beta}}} t^{1+\frac{1}{\beta}} \cdot \frac{1}{\beta \alpha^{1/\beta}} \cdot t^{\frac{1}{\beta}-1} e^{-t}\,dt \\
&= \frac{\alpha\beta}{\alpha^{1+\frac{1}{\beta}} \cdot \alpha^{\frac{1}{\beta}}\beta} \int_0^\infty t^{1+\frac{1}{\beta}} \cdot t^{\frac{1}{\beta}-1} e^{-t}\,dt \\
&= \frac{\alpha\beta}{\alpha \cdot \alpha^{\frac{1}{\beta}} \cdot \alpha^{\frac{1}{\beta}}\beta} \int_0^\infty t^{2/\beta} e^{-t}\,dt \\
&= \frac{1}{\alpha^{2/\beta}} \Gamma\left(1 + \frac{2}{\beta}\right) \\
&= \alpha^{-\frac{2}{\beta}} \Gamma\left(1 + \frac{2}{\beta}\right) \\
\therefore Var(X) &= E(X^2) - [E(X)] \\
&= \alpha^{-2/3} \Gamma\left(1 + \frac{2}{\beta}\right) - \left[\alpha^{-1/\beta}\Gamma\left(1+\frac{1}{\beta}\right)\right]^2 \\
&= \alpha^{-2/\beta} \Gamma\left(1 + \frac{2}{\beta}\right) - \alpha^{-2/\beta}\Gamma^2\left(1+\frac{1}{\beta}\right) \\
&= \alpha^{-2/\beta} \left[\Gamma\left(1+\frac{2}{\beta}\right) - \Gamma^2\left(1+\frac{1}{\beta}\right)\right]
\end{aligned}
$$

(the variance of Weibull distribution)

(7) **Rayleigh distribution**
A random variable X has a Rayleigh distribution if and only if its probability density function is given by:

$$f(x) = \begin{cases} 2\alpha x e^{-\alpha x^2} & , \text{ for } x > 0 \\ 0 & , \text{ elsewhere} \end{cases}$$

Show that the mean and variance of a random variable with Rayleigh distribution are respectively given by:

(i) $\mu = \dfrac{1}{2}\sqrt{\dfrac{\pi}{\alpha}}$ (ii) $\sigma^2 = \dfrac{1}{\alpha}\left(1 - \dfrac{\pi}{4}\right)$.

Proof (i)
Let X represent the random variable with Rayleigh distribution

$$\therefore \text{ Mean }(\mu) = E(X) = \int x f(x) dx$$
$$= \int_0^\infty x(2\alpha x e^{-\alpha x^2}) dx$$
$$= 2\int_0^\infty \alpha x^2 e^{-\alpha x^2} dx$$

Put $t = \alpha x^2$,

$$\therefore x = \left(\frac{t}{\alpha}\right)^{\frac{1}{2}} = \frac{1}{\alpha^{\frac{1}{2}}} t^{\frac{1}{2}}$$

$$\therefore dx = \frac{\frac{1}{2} t^{-1/2}}{\alpha^{1/2}} dt = \frac{t^{-1/2}}{2\alpha^{1/2}} dt$$

Using the above transformation,

$$\begin{aligned}
E(X) &= 2\int_0^\infty te^{-t}\frac{t^{-1/2}}{2\alpha^{1/2}}dt \\
&= \frac{2}{2\alpha^{1/2}}\int_0^\infty t\cdot t^{-1/2}e^{-t}dt \\
&= \frac{1}{\alpha^{1/2}}\int_0^\infty t^{1/2}e^{-t}dt \\
&= \frac{1}{\alpha^{1/2}}\Gamma(\tfrac{3}{2}) \\
&= \frac{1}{\alpha^{1/2}}\frac{1}{2}\cdot\Gamma(\tfrac{1}{2}) \\
&= \frac{1}{\alpha^{1/2}}\cdot\frac{1}{2}\sqrt{\pi} \\
&= \frac{1}{2}\frac{\sqrt{\pi}}{\sqrt{\alpha}} = \frac{1}{2}\sqrt{\frac{\pi}{\alpha}}.
\end{aligned}$$

Hence the mean (μ) of Rayleigh distribution $= \dfrac{1}{2}\sqrt{\dfrac{\pi}{\alpha}}$ proved.

Now,

$$\begin{aligned}
E(X^2) = \int x^2 f(x)dx &= \int_0^\infty x^2(2\alpha x e^{-\alpha x^2})dx \\
&= 2\alpha\int_0^\infty x^3 e^{-\alpha x^2}dx \\
&= 2\alpha\int_0^\infty \left[\left(\frac{t}{\alpha}\right)^{\frac{1}{2}}\right]^3 \cdot e^{-t}\frac{t^{-1/2}}{2\alpha^{1/2}}dt \\
&= 2\alpha\int_0^\infty \frac{t^{3/2}}{\alpha^{3/2}}\cdot\frac{t^{-1/2}}{2\alpha^{1/2}}e^{-t}dt \\
&= \frac{2\alpha}{2\alpha^2}\int_0^\infty t e^{-t}dt \\
&= \frac{1}{\alpha}\Gamma_2 = \frac{1}{\alpha}1\cdot\Gamma(1) = \frac{1}{\alpha}.
\end{aligned}$$

But,

$$\begin{aligned}Var(X) &= E(X^2) - [E(X)]^2 \\ &= \frac{1}{\alpha} - \left[\frac{1}{2}\sqrt{\frac{\pi}{\alpha}}\right]^2 \\ &= \frac{1}{\alpha} - \frac{1}{4}\left(\frac{\pi}{\alpha}\right) \\ &= \frac{1}{\alpha}\left(1 - \frac{\pi}{4}\right)\end{aligned}$$

Hence the variance (σ^2) of a random variable with Rayleigh distribution is $\dfrac{1}{\alpha}\left(1 - \dfrac{\pi}{4}\right)$ proved.

Joint Density Functions of Two Continuous Random Variables

Definition
If X and Y are two continuous random variables, $f(x,y)$ is said to be their joint probability density function if its satisfies the following conditions:

(i) $f(x,y) \geq 0$

(ii) $\displaystyle\int_{-\infty}^{\infty}\int_{-\infty}^{\infty} f(x,y)dxdy = 1.$

Examples.

(1) The joint density function of two continuous random variables X and Y is

$$f(x,y) = \begin{cases} kxy &, 0 \leq x \leq 3,\ 1 \leq y \leq 4 \\ 0 &, \text{elsewhere} \end{cases}$$

Find

(a) the value of the constant k

(b) find $P(1 \leq X \leq 3,\ 0 \leq Y \leq 3)$

(c) $P(X \geq 1,\ Y \leq 2)$.

Solution

(a) Since $f(x,y)$ is a joint probability density function

$$\therefore \int_0^3 \int_1^4 kxy\,dx\,dy = 1$$

$$\therefore k \int_{x=0}^{3} \left[\int_{y=1}^{4} (xy\,dy) \right] dx = 1.$$

$$\therefore k \int_{k=0}^{3} \left. \frac{xy^2}{2} \right|_1^4 = 1$$

$$\therefore k \int_{x=0}^{3} \frac{x}{2}(16-1)dx = 1$$

$$\therefore \frac{15k}{2} \int_{x=0}^{3} x\,dx = 1$$

$$\therefore \frac{15k}{2} \left. \frac{x^2}{2} \right|_0^3 = 1$$

$$\therefore \frac{135k}{4} = 1$$

$$\therefore k = \frac{4}{135}$$

Hence,

$$f(x,y) = \begin{cases} \dfrac{4}{135}xy & 0 < x < 3,\ 1 < y < 4 \\ 0 & \text{elsewhere} \end{cases}$$

(b) $\quad P(1 \leq X \leq 3,\ 0 \leq Y \leq 3);$
$$= P(1 \leq X \leq 3,\ 1 \leq Y \leq 3) \text{ (within the range given)}$$
$$= \int_{x=1}^{3} \int_{y=1}^{3} f(x,y)\,dx\,dy$$
$$= \int_{x=1}^{3} \int_{y=1}^{3} \frac{4}{135} xy\,dx\,dy$$
$$= \frac{4}{135} \int_{x=1}^{3} \left[\int_{y=1}^{3} (xy)\,dy \right] dx$$
$$= \frac{4}{135} \int_{x=1}^{3} \left[\frac{x}{2}(y^2) \Big|_{y=1}^{3} \right] dx$$
$$= \frac{4}{135} \int_{x=1}^{3} 4x\,dx$$
$$= \frac{4}{135} 2x^2 \Big|_{1}^{3}$$
$$= \frac{8}{135}(9-1) = \frac{64}{135}$$

$\therefore\ P(1 \leq X \leq 3,\ 0 \leq Y \leq 3) = \dfrac{64}{135}$

(c) $\quad P(X \geq 1,\ Y \leq 2) = P(1 \leq X \leq 3,\ 1 \leq Y \leq 2)$
$$= \int_{x=1}^{3} \int_{y=1}^{2} \frac{4}{135} xy\,dx\,dy \text{ (within the range given)}$$
$$= \frac{4}{135} \int_{x=1}^{3} \frac{xy^2}{2} \Big|_{y=1}^{2} dx$$
$$= \frac{4}{135} \int_{x=1}^{3} \frac{3x}{2}\,dx$$
$$= \frac{1}{45} x^2 \Big|_{1}^{3} = \frac{8}{45}$$

$\therefore\ P(X \geq 1,\ Y \leq 2) = \dfrac{8}{45}.$

Chapter 7

(2) Given that the joint density function of two continuous random variables is

$$f(x,y) = \begin{cases} t(x+2y) &, 1 \le x \le 3, \ 0 \le y \le 2 \\ 0 &, \text{elsewhere} \end{cases}$$

Find

(a) the constant t

(b) $P(2 < X < 5, \ Y \ge 1)$

Solution

(a) Since $f(x,y)$ is a probability density function

$$\therefore \int_{x=1}^{3} \int_{y=0}^{2} t(x+2y)\,dy\,dx = 1.$$

$$\therefore t \int_{x=1}^{3} \left[\int_{y=0}^{2} (x+2y)\,dy \right] dx = 1$$

$$\therefore t \int_{x=1}^{3} (xy+y^2) \Big|_{0}^{2} dx = 1$$

$$\therefore t \int_{x=1}^{3} (2x+4)\,dx = 1$$

$$\therefore t(x^2+4x)\big|_{1}^{3} = 1$$

$$\therefore t(21-5) = 1$$

$$\therefore 16t = 1 \Rightarrow t = \frac{1}{16}$$

Hence

$$f(x,y) = \begin{cases} \dfrac{1}{16}(x+2y) &, 1 \le x \le 3, \ 0 \le y \le 2 \\ 0 &, \text{elsewhere} \end{cases}$$

(b) $P(2 < X < 5, Y \geq 1)$ = $P(2 < X \leq 3, 1 \leq Y \leq 2)$
(within the range given above)

$$= \int_{x=2}^{3} \int_{y=1}^{2} \frac{1}{16}(x+2y)dxdy$$

$$= \frac{1}{16} \int_{x=2}^{3} \left[\int_{y=1}^{2}(x+2y)dy\right] dx$$

$$= \frac{1}{16} \int_{x=2}^{3} (xy+y^2)\Big|_{1}^{2} dx$$

$$= \frac{1}{16} \int_{x=2}^{3} [(2x+4)-(x+1)]dx$$

$$= \frac{1}{16} \int_{x=2}^{3} (x+3)dx$$

$$= \frac{1}{16}\left(\frac{x^2}{2}+3x\right)\Big|_{2}^{3}$$

$$= \frac{1}{16}\left[\left(\frac{9}{2}+9\right)-(2+6)\right]$$

$$= \frac{1}{16}\left[\frac{11}{2}\right] = \frac{11}{32}$$

Hence = $P(2 < X < 5, Y \geq 1) = \dfrac{11}{32}$

Conditional Probability Function
Definition

If X and Y are two continuous random variables with joint probability density function $f(x,y)$, $f(y/x)$ is said to be the conditional probability function of Y given X where $f(y/x) = \dfrac{f(x,y)}{f_1(x)}$ and $f_1(x) = \int f(x,y)dy$ is the marginal probability density function of X.

Similarly, the conditional probability function of X given Y is given by $f(x/y) = \dfrac{f(x,y)}{f_2(y)}$ where $f_2(y) = \int f(x,y)dx$ is the marginal probability density function of Y.

Example 1

(1) The joint density function of the random variables X and Y is given by

$$f(x,y) = \begin{cases} bxy & ,0 \leq x \leq 1, \ 0 \leq y \leq 1 \\ 0 & ,\text{elsewhere} \end{cases}$$

Find

(i) the constant b

(ii) the marginal probability density function of Y

(iii) the marginal probability density function of X

(iv) the conditional probability density function of X given Y

(v) the conditional probability density function of Y given X.

Solution

(i) Since $f(x,y)$ is a joint density function,

$$\int_{x=0}^{1} \int_{y=0}^{1} bxy\,dx\,dy = 1$$

$$\therefore b \int_{x=0}^{1} \int_{y=0}^{1} (xy\,dy)\,dx = 1$$

$$\therefore b \int_{x=0}^{1} \left(\frac{xy^2}{2}\Big|_0^1\right) dx = 1$$

$$\therefore b \int_{x=0}^{1} \frac{x}{2}\,dx = 1$$

$$\therefore b\frac{x^2}{4}\Big|_0^1 = 1$$

$$\therefore \frac{b}{4} = 1$$

$$\therefore b = 4$$

Hence,
$$f(x,y) = \begin{cases} 4xy, & 0 \leq x \leq 1,\ 0 \leq y \leq 1 \\ 0, & \text{elsewhere} \end{cases}$$

(ii) The marginal probability density function of X,

$$\begin{aligned}
&= f_1(x) \\
&= \int_0^1 f(x,y)\,dy \\
&= \int_0^1 4xy\,dy \\
&= 2xy^2 \Big|_{y=0}^{1} \\
\therefore f_1(x) &= 2x.
\end{aligned}$$

(iii) The marginal probability density function of Y,

$$\begin{aligned}
&= f_2(y) \\
&= \int_0^1 4xy\,dx \\
&= 2x^2 y \Big|_{x=0}^{1} \\
\therefore f_2(y) &= 2y.
\end{aligned}$$

(iv) The conditional probability density function of X given Y is

$$f(x/y) = \frac{f(x,y)}{f_2(y)} = \frac{4xy}{2y} = 2x$$
$$\therefore f(x/y) = 2x$$

(v) The conditional probability density function of Y given X is

$$f(x/y) = \frac{f(x,y)}{f_1(x)} = \frac{4xy}{2x}$$
$$\therefore f(x/y) = 2y$$

Example 2
The joint density function of the random variables X and Y is given by

$$f(x,y) = \begin{cases} \dfrac{1}{4}(2x+y) & ,0 \leq x \leq 1,\ 0 \leq y \leq 2 \\ 0 & ,\text{otherwise} \end{cases}$$

Find the conditional density function of Y given X.

Solution
The conditional density function of Y given X is

$$f(y/x) = \frac{f(x,y)}{f_1(x)}$$

$$\begin{aligned}
f_1(x) &= \int_0^2 \frac{1}{4}(2x+y)dy \\
&= \frac{1}{4}\left(2xy + \frac{y^2}{2}\right)\Bigg|_0^2 \\
&= \frac{1}{4}(4x+2) \\
&= \frac{1}{2}(2x+1).
\end{aligned}$$

Hence,

$$f(y/x) = \frac{f(x,y)}{f_1(x)} = \frac{\dfrac{1}{4}(2x+y)}{\dfrac{1}{2}(2x+1)} = \frac{1}{2}\left(\frac{2x+y}{2x+1}\right)$$

$$\therefore\ f(y/x) = \frac{2x+y}{2(2x+1)}$$

Expectation and Variance of Joint Distributions
If X and Y are two continuous random variables having joint density function $f(x,y)$, the means or expectations of X and Y are given respectively by

Chapter 7

(i) $\mu_x = E(X) = \int_{-\infty}^{\infty} \int_{-\infty}^{\infty} x f(x,y) dx dy$

(ii) $\mu_y = E(Y) = \int_{-\infty}^{\infty} \int_{-\infty}^{\infty} y f(x,y) dx dy$

The variance of X and Y are also given by

$$\sigma_x^2 = Var(X) = E[(X - \mu_x)^2]$$
$$= \int_{-\infty}^{\infty} \int_{-\infty}^{\infty} (x - \mu_x)^2 f(x,y) dx dy$$
$$\sigma_y^2 = Var(Y) = E[(Y - \mu_y)^2]$$
$$= \int_{-\infty}^{\infty} \int_{-\infty}^{\infty} (y - \mu_y)^2 f(x,y) dx dy$$

Example

If X and Y are two continuous random variables with joint density function

$$f(x,y) = \begin{cases} \dfrac{1}{16}(x+2y) &, 1 \leq x \leq 3,\ 0 \leq y \leq 2 \\ 0 &, \text{elsewhere} \end{cases}$$

Find the expectation and variance of X.

Solution

$$\begin{aligned}
E(X) &= \int_{-\infty}^{\infty} \int_{-\infty}^{\infty} x f(x,y) dx dy \\
&= \int_0^2 \int_1^3 x \left[\frac{1}{16}(x+2y)\right] dx dy \\
&= \frac{1}{16} \int_0^2 \int_1^3 [x(x+2y)] dx dy \\
&= \frac{1}{16} \int_0^2 \int_1^3 (x^2 + 2xy) dx dy \\
&= \frac{1}{16} \int_0^2 \left|\frac{x^3}{3} + x^2 y\right|_1^3 dy
\end{aligned}$$

$$= \frac{1}{16}\int_0^2 \left[9+9y-\frac{1}{3}-y\right]dy$$

$$= \frac{1}{16}\int_0^2 \left(\frac{26}{3}+8y\right)dy$$

$$= \frac{1}{16}\left|\frac{26y}{3}+4y^2\right|_0^2$$

$$= \frac{1}{16}\left[\frac{52}{3}+16\right]$$

$$= \frac{1}{16}\left[\frac{100}{3}\right] = \frac{25}{12} = 2\frac{1}{12}$$

Hence $E(X) = 2\frac{1}{12}$

$$E(X^2) = \int_0^2\int_1^3 x^2 f(x,y)dxdy$$

$$= \int_0^2\int_1^3 x^2\left[\frac{1}{16}(x+2y)\right]dxdy$$

$$= \frac{1}{16}\int_0^2\int_1^3 \left[(x^3+2x^2y)\right]dxdy$$

$$= \frac{1}{16}\int_0^2 \left|\frac{x^4}{4}+\frac{2x^3y}{3}\right|_1^3 dy$$

$$= \frac{1}{16}\int_0^2 \left\{\left(\frac{81}{4}+18y\right)-\left(\frac{1}{4}+\frac{2}{3}y\right)\right\}dy$$

$$= \frac{1}{16}\int_0^2 \left(20+\frac{52y}{3}\right)dy$$

$$= \frac{1}{16}\left(20y+\frac{52y^2}{6}\bigg|_0^2\right)$$

$$= \frac{1}{16}\left(40+\frac{208}{6}\right)$$

$$= \frac{1}{16}\left(\frac{448}{6}\right) = \frac{14}{3} = 4\frac{2}{3}$$

Chapter 7

Hence

$$\begin{aligned} Var(X) &= E(X^2) - \mu_x^2 \\ &= \frac{14}{3} - \left(\frac{25}{12}\right)^2 \\ &= \frac{14}{3} - \frac{625}{144} \\ &= \frac{672 - 625}{144} = \frac{47}{144} \\ Var(X) &= \frac{47}{144} \end{aligned}$$

Conditional Expectation or Conditional Mean
Definition
If X and Y have joint density function $f(x,y)$, the conditional expectation or conditional mean of Y given X is given by

$$E(Y/X) = \int_{-\infty}^{\infty} y f(y/x) dy \qquad \text{where } f(y/x) = \frac{f(x,y)}{f_1(x)}$$

The conditional expectation or conditional mean of X given Y is

$$E(X/Y) = \int_{-\infty}^{\infty} x f(x/y) dx \qquad \text{where } f(x/y) = \frac{f(x,y)}{f_2(y)}$$

Note: $E(Y/X)$ is the same as the regression curve of Y on X and $E(X/Y)$ is the regression curve of X on Y.

Example 1
If X and Y are random variables with joint density function

$$f(x,y) = \begin{cases} \frac{1}{7}(3x + 2y) & 0 \leq x \leq 1,\ 0 \leq y \leq 2 \\ 0 & \text{elsewhere} \end{cases}$$

Find the conditional expectation of X given Y.

Chapter 7

Solution

$$f_2(y) = \int_0^1 f(x,y)dx$$

$$= \int_0^1 \frac{1}{7}(3x+2y)dx$$

$$= \frac{1}{7}\left[\frac{3}{2}x^2 + 2xy\right]_0^1$$

$$= \frac{1}{7}\left[\frac{3}{x} + 2y\right] = \frac{3+4y}{14}$$

$$f(x/y) = \frac{f(x,y)}{f_2(y)} = \frac{\frac{1}{7}(3x+2y)}{\frac{3+4y}{14}}$$

$$= \frac{\frac{1}{7}(3x+2y) \times 14}{3+4y}$$

$$= \frac{2(3x+2y)}{3+4y} = \frac{6x+4y}{3+4y}$$

$$E(X/Y) = \int_0^1 xf(x/y)dx = \int_0^1 x\frac{(6x+4y)}{3+4y}dx$$

$$= \int_0^1 \frac{1}{3+4y}(6x^2+4xy)dx$$

$$= \frac{1}{3+4y}\left[2x^3 + 2x^2y\right]_0^1$$

$$= \frac{1}{3+4y}[2+2y]$$

$$= \frac{2(1+y)}{3+4y}$$

Example 2
The joint probability density function of the random variables X and Y is

Chapter 7

given by
$$f(x,y) = \begin{cases} \frac{1}{4}(2x+y) & 0 \le x \le 1, \ 0 \le y \le 2 \\ 0 & \text{otherwise} \end{cases}$$

Find the regression curve of Y on X.

Solution

The regression curve of Y on X is the same as $E(Y/X)$. But

$$E(Y/X) = \int_{-\infty}^{\infty} y f(y/x) dy = \int_{-\infty}^{\infty} \frac{y f(x,y)}{f_1(x)} dy$$

For the given density function

$$E(Y/X) = \int_0^2 \frac{y(\frac{1}{4}(2x+y))}{f_1(x)} dy$$

Now,

$$\begin{aligned} f_1(x) &= \int_0^2 f(x,y) dy = \int_0^2 \frac{1}{4}(2x+y) dy \\ &= \frac{1}{4}\left(2xy + \frac{1}{2}y^2\right)\bigg|_0^2 \\ &= \frac{1}{4}(4x+2) \\ &= \frac{1}{2}(2x+1) \end{aligned}$$

Hence,

$$f(y/x) = \frac{f(x,y)}{f_1(x)} = \frac{\frac{1}{4}(2x+y)}{\frac{1}{2}(2x+1)}$$

$$= \frac{1}{2}\left(\frac{2x+y}{2x+1}\right)$$

$$\therefore E(Y/X) = \int_0^2 y\left[\frac{1}{2}\left(\frac{2x+y}{2x+1}\right)\right] dy$$

$$= \frac{1}{2(2x+1)} \int_0^2 y(2x+y) dy$$

$$= \frac{1}{2(2x+1)} \int_0^2 (2xy+y^2) dy$$

$$= \frac{1}{2(2x+1)} \left(xy^2 + \frac{1}{3}y^3\right)\Big|_0^2$$

$$= \frac{1}{2(2x+1)} \left(4x + \frac{8}{3}\right)$$

$$= \frac{(12x+8)}{6(2x+1)} = \frac{2(3x+2)}{3(2x+1)}$$

Hence the regression curve of Y on X is $\frac{2(3x+2)}{3(2x+1)}$.

Conditional Variance
Definition

If X and Y have joint probability density function $f(x,y)$, the conditional variance of X given Y is defined as

$$E\left[(X-\mu_x)^2/Y\right] = \int_{-\infty}^{\infty} (x-\mu_x)^2 f(x/y) dx$$

where $\mu_x = E(X/Y)$ and the conditional variance of Y given X is also defined as

$$E\left[(Y-\mu_y)^2/X\right] = \int_{-\infty}^{\infty} (y-\mu_y)^2 f(y/x) dx$$

Chapter 7

where $\mu_y = E(Y/X)$.
Note:
$$E\left[(X-\mu_x)^2/Y\right] = E(X^2/Y) - [E(X/Y)]^2$$
$$E\left[(Y-\mu_y)^2/X\right] = E(Y^2/X) - [E(Y/X)]^2$$

Example
Find the variance of Y given X of the distribution given in example 2 under conditional expectation above.

Solution
The joint density function given in example 2 under the conditional expectation is

$$f(x,y) = \begin{cases} \dfrac{1}{4}(2x+y) & , \quad 0 \le x \le 1, \; 0 \le y \le 3 \\ 0 & , \quad \text{elsewhere} \end{cases}$$

For the given distribution

$$E(Y/X) = \frac{2(3x+2)}{3(2x+1)} \quad \text{(from example 2)}$$

Now,

$$\begin{aligned}
E(Y^2/X) &= \int_0^2 y^2 f(y/x) dy \quad \text{where } f(y/x) = \frac{2x+y}{2(2x+1)} \\
&= \frac{1}{2(2x+1)} \int_0^2 y^2(2x+y) dy \\
&= \frac{1}{2(2x+1)} \int_0^2 (2xy^2 + y^3) dy \\
&= \frac{1}{2(2x+1)} \left[\frac{2}{3}xy^3 + \frac{y^4}{4}\right]_0^2 \\
&= \frac{1}{2(2x+1)} \left(\frac{16x}{3} + 4\right) \\
&= \frac{2(4x+3)}{3(2x+1)}
\end{aligned}$$

Chapter 7

Hence,

$$\begin{aligned}
Var(Y/X) &= E\left(Y^2/X\right) - [E\left(Y/X\right)]^2 \\
&= \frac{2(4x+3)}{3(2x+1)} - \left\{\frac{2(3x+2)}{3(2x+1)}\right\}^2 \\
&= \frac{2}{3}\left(\frac{4x+3}{2x+1}\right) - \frac{4}{9}\frac{(3x+2)^2}{(2x+1)^2} \\
&= \frac{6(4x+3)(2x+1) - 4(3x+2)^2}{9(2x+1)^2} \\
&= \frac{12x^2 + 12x + 2}{9(2x+1)^2} \\
&= \frac{2(6x^2 + 12x + 1)}{9(2x+1)^2}.
\end{aligned}$$

Chapter 7

Moment Generating Function (m.g.f) And Cummulant Generating Function

Moment Generating Function

Definition
The moment generating function of a random variable X is defined as $M_X(\theta) = E(e^{\theta x})$. For discrete random variable,

$$M_X(\theta) = \sum_{j=1}^{n} e^{\theta x_j} f(x_j) = \sum e^{\theta x} f(x)$$
$$= E(e^{\theta x}).$$

For a continuous random variable,

$$M_X(\theta) = \int_{-\infty}^{\infty} e^{\theta x} f(x) dx$$
$$= E(e^{\theta x}).$$

Recall that
$$e^x = 1 + x + \frac{x^2}{2!} + \frac{x^3}{3!} + \cdots + \frac{x^r}{r!} + \cdots$$

Chapter 7

$$M_X(\theta) = E(e^{\theta x}) = E\left[1 + \theta x + \frac{(\theta x)^2}{2!} + \frac{(\theta x)^3}{3!} + \cdots + \frac{(\theta x)^r}{r!} + \cdots\right]$$

$$= 1 + \theta E(x) + \frac{\theta^2 E(x^2)}{2!} + \frac{\theta^2 E(x^3)}{3!} + \cdots + \frac{\theta^r E(x^r)}{r!} + \cdots$$

$$= 1 + \theta \mu'_1 + \frac{\theta^2}{2!}\mu'_2 + \frac{\theta^3}{3!}\mu'_3 + \cdots + \frac{\theta^r}{r!}\mu'_r + \cdots$$

where $\mu'_1, \mu'_2, \ldots, \mu'_r$ are the 1st, 2nd, 3rd, ... rth moments about the origin. The coefficient in the above expansion gives us various moments about the origin, hence the name moment generating function.

Example 1

The random variable X can assume the values 2 and -2 with probability $\dfrac{1}{2}$ each. Find (a) the moment generating function and (b) the first four moments about the origin from the moment generating function obtained.

Solution

Let $x_1 = 2$ and $x_2 = -2$.
We know that

$$M_X(t) = E(e^{tx}) = \sum_{t=1}^{2} e^{tx_i} f(x) = e^{tx_1} f(x_1) + e^{tx_2} f(x_2)$$

$$= e^{2t}\left(\frac{1}{2}\right) + e^{-2t}\left(\frac{1}{2}\right)$$

$$= \frac{1}{2}\left[e^{2t} + e^{-2t}\right]$$

$$e^{2t} = 1 + 2t + \frac{(2t)^2}{2!} + \frac{(2t)^3}{3!} + \frac{(2t)^4}{4!} + \cdots + \frac{(2t)^r}{r!} + \cdots$$

$$= 1 + 2t + \frac{4t^2}{2!} + \frac{8t^3}{3!} + \frac{16t^4}{4!} + \cdots + \frac{2^r t^r}{r!} + \cdots$$

$$e^{-2t} = 1 - 2t + \frac{4t^2}{2!} - \frac{8t^3}{3!} + \frac{16t^4}{4!} + \cdots - \frac{2^r t^r}{r!} + \cdots$$

$$\therefore \quad e^{2t} + e^{-2t} = 2 + 2\left(\frac{4t^2}{2!}\right) + 2\left(\frac{16t^4}{4!}\right) + \cdots$$

Chapter 7   160

$$M_X(t) = \frac{1}{2}(e^{2t} + e^{-2t}) = 1 + \left(\frac{4t^2}{2!}\right) + \left(\frac{16t^4}{4!}\right) + \cdots$$

$$\therefore \quad M_X(t) = 1 + 4\left(\frac{t^2}{2!}\right) + 16\left(\frac{t^4}{4!}\right) + \cdots \tag{1}$$

Recall that

$$M_X(t) = 1 + \mu_1' t + \mu_2' \frac{t^2}{2!} + \mu_3' \frac{t^3}{3!} + \mu_4' \frac{t^4}{4!} + \cdots \tag{2}$$

From (1) and (2), $\mu_1' = 0$, $\mu_2' = 4$, $\mu_3' = 0$ and $\mu_4' = 16$ which are the required first four moments about the origin.

Example 2

A random variable X has density function given by

$$f(x) = \begin{cases} 3e^{-3x} & , \ x \geq 0 \\ 0 & , \ \text{otherwise} \end{cases}$$

find (a) the moment generating function of the above and use it to generate the first five moments about the origin.

Solution

$$\begin{aligned}
\text{(a)} \quad M_X(\theta) = E(e^{\theta x}) &= \int_{-\infty}^{\infty} e^{\theta x} f(x) dx \\
&= \int_0^{\infty} e^{\theta x}(3e^{-3x}) dx \\
&= 3 \int_0^{\infty} e^{-x(3-\theta)} dx \\
&= \left. \frac{3e^{-x(3-\theta)}}{\theta - 3} \right|_0^{\infty} \\
&= \frac{-3}{\theta - 3} \\
&= \frac{3}{3 - \theta}
\end{aligned}$$

Chapter 7

If $|\theta| < 3$, then

$$M_X(\theta) = \frac{3}{3-\theta} = \frac{1}{1-\frac{\theta}{3}} = 1 + \frac{\theta}{3} + \frac{\theta^2}{9} + \frac{\theta^3}{27} + \frac{\theta^4}{81} + \frac{\theta^5}{243} + \cdots \quad (1)$$

But,

$$M_X(\theta) = 1 + \mu'_1\theta + \mu'_2\frac{\theta^2}{2!} + \mu'_3\frac{\theta^3}{3!} + \mu'_4\frac{\theta^4}{4!} + \cdots + \mu_r\frac{\theta^5}{r!} + \cdots \quad (2)$$

Comparing (1) and (2), we have

$$\mu'_1 = \mu = \frac{1}{3}, \quad \frac{\mu'_2}{2!} = \frac{1}{9} \Rightarrow \mu'_2 = \frac{2}{9}$$

$$\frac{\mu'_3}{3!} = \frac{1}{27} \Rightarrow \mu'_3 = \frac{3!}{27} = \frac{2}{9}$$

$$\frac{\mu'_4}{4!} = \frac{1}{81} \Rightarrow \mu'_4 = \frac{4!}{81} = \frac{8}{27}$$

$$\frac{\mu'_5}{5!} = \frac{1}{243} \Rightarrow \mu'_5 = \frac{5!}{243} = \frac{40}{81}$$

Hence, the first five moments about the origin are
$\mu = \frac{1}{3},\ \mu'_2 = \frac{2}{9},\ \mu'_3 = \frac{2}{9},\ \mu'_4 = \frac{8}{27},\ \mu'_5 = \frac{40}{81}$

Properties of Moment Generating Function $M_X(\theta)$

Recall that $M_X(\theta) = 1 + \theta\mu'_1 + \frac{\theta^2}{2!}\mu'_2 + \frac{\theta^3}{3!}\mu'_3 + \cdots + \frac{\theta^r}{r!}\mu'_r + \cdots$.

$$\begin{aligned}
\text{(i)} \quad \therefore\ M'_X(\theta) &= \frac{dM(\theta)}{d\theta} \\
&= \left(\mu'_1 + \frac{2\theta\mu'_2}{2!} + \frac{3\theta^2\mu'_3}{3!} + \frac{4\theta^3\mu'_4}{4!} + \cdots\right) \\
\therefore\ M'_X(\theta) &= \left(\mu_1 + \theta\mu'_2 + \frac{3\theta^2\mu'_3}{3!} + \frac{4\theta^3\mu'_4}{4!} + \cdots\right) \quad (1)\\
\therefore\ M'_X(0) &= \mu'_1 = \mu = E(X)
\end{aligned}$$

Chapter 7

From (i) above,

$$
\begin{aligned}
\text{(ii)} \quad M_X''(\theta) &= \frac{d^2 M(\theta)}{d\theta^2} \\
&= \left(\mu_2' + \frac{6\theta \mu_3'}{3!} + \frac{12\theta^2 \mu_4'}{4!} + \cdots\right) \\
\therefore M_X''(0) &= \mu_2' = E(X^2) \qquad (2) \\
\text{(iii)} \quad M_X'''(\theta) &= \frac{6\mu_3'}{3!} + \frac{24\theta \mu_4'}{4!} + \cdots \\
&= \mu_3' + \theta \mu_4' + \cdots \\
M_X'''(0) &= \mu_3'
\end{aligned}
$$

In general, $M_X^r(0) = \mu_r'$ where μ_r' is the rth moment about the origin.
From above,

$$
\begin{aligned}
Var(X) &= E(X^2) - [E(X)]^2 \\
&= \mu_2' - (\mu_1')^2 \\
&= M_X''(0) - [M_X'(0)]^2
\end{aligned}
$$

Determination of the mean and variance of some probability distributions using property of moment generating function.

(1) Binomial Distribution

$$f(x) = \binom{n}{x} p^x q^{n-x}$$

$$
\begin{aligned}
M(\theta) = E(e^{\theta x}) &= \sum_{x=0}^{n} e^{\theta x} \binom{n}{x} p^x q^{n-x} \\
&= \sum_{x=0}^{n} \binom{n}{x} (pe^\theta)^x q^{n-x} \\
&= (pe^\theta + q)^n \quad \text{(by Binomial expansion)}
\end{aligned}
$$

Chapter 7

We can use properties (1) and (2) to find the mean and variance of the resulting moment generating function.

$$\begin{aligned}
M'_X(\theta) &= npe^\theta(pe^\theta + q)^{n-1} \\
M'_X(0) &= np(p+q)^{n-1} = np \\
M'_X(0) &= E(X) = np \\
M''_X(\theta) &= npe^\theta(pe^\theta + q)^{n-2}(n-1)pe^\theta + npe^\theta(pe^\theta + q)^{n-1} \\
M''_X(0) &= np(p+q)^{n-2}(n-1)p + np(p+q)^{n-1} \\
&= np(n-1)p + np \\
&= n(n-1)p^2 + np
\end{aligned}$$

$$\begin{aligned}
\text{Variance } (X) &= M''_X(0) - [M'_X(0)]^2 \\
&= n(n-1)p^2 + np - (np)^2 \\
&= n^2p^2 - np^2 + np - n^2p^2 \\
&= np - np^2 = np(1-p) = npq \quad (\because p+q=1)
\end{aligned}$$

(2) Poisson Distribution

$$f(x) = \frac{\lambda^x e^{-\lambda}}{x!}$$

Moment Generating Function,

$$\begin{aligned}
M_X(\theta) &= E(e^{\theta x}) \\
&= \sum_{x=0}^{\infty} e^{\theta x} f(x) \\
&= \sum_{x=0}^{\infty} e^{\theta x} \cdot \frac{\lambda^x e^{-\lambda}}{x!} \\
&= e^{-\lambda} \sum_{x=0}^{\infty} \frac{(e^\theta \lambda)^x}{x!} \\
&= e^{-\lambda} \sum_{x=0}^{\infty} \frac{(\lambda e^\theta)^x}{x!}
\end{aligned}$$

$$
\begin{aligned}
&= e^{-\lambda}\left[1 + \lambda e^{\theta} + \frac{(\lambda e^{\theta})^2}{2!} + \frac{(\lambda e^{\theta})^3}{3!} + \cdots\right] \\
&= e^{-\lambda}e^{\lambda e^{\theta}} \\
\therefore M_X(\theta) &= e^{\lambda(e^{\theta}-1)} \\
M'_X(\theta) &= \lambda e^{\theta} e^{\lambda(e^{\theta}-1)} \\
M'_X(0) &= \lambda e^{0} e^{\lambda(0)} = \lambda \\
\therefore \text{Mean} &= M'_X(0) = \lambda \\
M''_X(\theta) &= \lambda e^{\theta} \cdot \lambda e^{\theta} e^{\lambda(e^{\theta}-1)} + \lambda e^{\theta} \cdot e^{\lambda(e^{\theta}-1)} \\
&= \lambda e^{\theta} e^{\lambda(e^{\theta}-1)}[\lambda e^{\theta} + 1] \\
\therefore M''_X(0) &= \lambda e^{0} e^{\lambda(e^{0}-1)}[\lambda e^{0} + 1] \\
&= \lambda[\lambda + 1] \\
&= \lambda^2 + \lambda \\
\therefore Var(X) &= M''_X(0) - [M'_X(0)]^2 \\
&= \lambda^2 + \lambda - (\lambda)^2 \\
&= \lambda.
\end{aligned}
$$

(3) Uniform or Rectangular Distribution.

A discrete random variable X is said to have a uniform distribution if and only if its probability density function is given by

$$
f(x) = \begin{cases} \dfrac{1}{k} & , \text{ for } x = 1, 2, \ldots, k \\ 0 & , \text{ elsewhere} \end{cases}
$$

where the parameter k ranges over positive integer and X is a discrete uniform random variable.

Its moment generating function is

$$M_X(\theta) = \sum_{x=1}^{k} e^{\theta x} f(x), \quad x = 1, 2, 3, \ldots, k$$

$$= \sum_{x=1}^{k} e^{\theta x} \left(\frac{1}{k}\right)$$

$$= \frac{1}{k} \sum_{x=1}^{k} e^{\theta x}$$

$$\therefore \quad M_X(\theta) = \frac{1}{k} \left\{ e^{\theta} + e^{2\theta} + e^{3\theta} + \cdots + e^{k\theta} \right\} \quad (1)$$

The expression in the bracket is a geometric series with e^{θ} as the first term and e^{θ} as the common ratio

$$\therefore \quad M_X(\theta) = \frac{1}{k} \left\{ \frac{e^{\theta}(1 - e^{\theta k})}{1 - e^{\theta}} \right\}$$

From (1),

$$M'_X(\theta) = \frac{1}{k} \{ e^{\theta} + 2e^{2\theta} + 3e^{3\theta} + \cdots + k e^{k\theta} \} \quad (2)$$

$$M'_X(0) = \frac{1}{k} \{ 1 + 2 + 3 + \cdots + k \}$$

The expression in the bracket is Arithmetic series whose first term is 1 and common difference is also 1. The last term is k.

$$\therefore \quad M'_X(0) = \frac{1}{k} \left\{ \frac{k}{2}[2(1) + (k-1)1] \right\}$$

$$= \frac{1}{k} \left\{ \frac{k}{2}[2 + k - 1] \right\}$$

$$= \frac{1}{k} \left\{ \frac{k}{2}(1 + k) \right\}$$

$$= \frac{k+1}{2}.$$

Chapter 7

∴ But $M'_X(0)$ is the mean of the distribution.
∴ Mean of the distribution $= \dfrac{k+1}{2}$.
From (2),

$$M''_X(\theta) = \frac{1}{k}\{e^\theta + 4e^{2\theta} + 9e^{3\theta} + \cdots + k^2 e^{k\theta}\}$$

$$M''_X(0) = \frac{1}{k}\{1 + 4 + 9 + \cdots + k^2\}$$

$$= \frac{1}{k}\left\{\frac{1}{6}k(k+1)(2k+1)\right\}$$

$$= \frac{(k+1)(2k+1)}{6}$$

∴ $Var(X) = M''_X(0) - [M'_X(0)]^2$

$$= \frac{(k+1)(2k+1)}{6} - \left(\frac{k+1}{2}\right)^2$$

$$= \frac{2k^2 + 3k + 1}{6} - \frac{(k^2 + 2k + 1)}{4}$$

$$= \frac{2(2k^2 + 3k + 1) - 3(k^2 + 2k + 1)}{12}$$

$$= \frac{k^2 - 1}{12}.$$

(4) Bernoulli distribution
A discrete random variable X is said to have a Bernoulli distribution if

$$f(x) = \begin{cases} p^x q^{1-x} &, \quad x = 0, 1, \text{ and } p + q = 1 \\ 0 &, \quad \text{elsewhere} \end{cases}$$

Chapter 7

$$\begin{aligned}
M_X(\theta) &= \sum e^{\theta x} f(x) \\
&= \sum e^{\theta x} p^x q^{1-x} \\
&= q \sum \left(\frac{e^\theta p}{q}\right)^x \quad x = 0, 1. \\
&= q \left[1 + e^\theta \left(\frac{p}{q}\right)\right] \\
&= q + pe^\theta.
\end{aligned} \quad (1)$$

From (1),
$$M'_X(\theta) = pe^\theta \quad (2)$$
$$\therefore \quad M'_X(0) = pe^0 = p.$$

∴ Mean of Bernoulli distribution is p.
From (2),

$$\begin{aligned}
M''_X(\theta) &= pe^\theta \\
\therefore \quad M''_X(0) &= pe^0 = p \\
\therefore \quad Var(X) &= M''_X(0) - [M'_X(0)]^2 \\
&= p - p^2 \\
&= p(1-p) = pq.
\end{aligned}$$

(5) Geometric Distribution

A random variable X is said to have a Geometric distribution if its probability density function

$$f(x) = \begin{cases} pq^{x-1}, & \text{for } x = 1, 2, 3, \ldots \\ 0, & \text{elsewhere} \end{cases}$$

Chapter 7

Moment Generating Function of a Geometric distribution is

$$\begin{aligned}
M_X(\theta) &= \sum e^{\theta x} f(x) \\
&= \sum e^{\theta x} pq^{x-1} \quad \text{for } 1,2,3\ldots \\
&= \frac{p}{q} \sum (e^\theta q)^x \\
&= \frac{p}{q}\{e^\theta q + e^{2\theta} q^2 + e^{3\theta} q^3 + \cdots\}
\end{aligned}$$

The expression in the bracket is an infinite geometric series having first term $e^\theta q$, common ratio $e^\theta q$.

$$\begin{aligned}
\therefore M_X(\theta) &= \frac{p}{q}\left\{\frac{qe^\theta}{1 - qe^\theta}\right\} \\
&= \frac{pe^\theta}{1 - qe^\theta} \quad (1) \\
M_X'(\theta) &= \frac{(1 - qe^\theta)pe^\theta - pe^\theta(-qe^\theta)}{(1 - qe^\theta)^2} \\
&= \frac{pe^\theta - pqe^{2\theta} + pqe^{2\theta}}{(1 - qe^\theta)^2} \\
&= \frac{pe^\theta}{(1 - qe^\theta)^2} \quad (2) \\
\therefore M_X'(0) &= \frac{pe^0}{(1 - qe^0)^2} = \frac{p}{(1 - q)^2} = \frac{p}{p^2} = \frac{1}{p}
\end{aligned}$$

Chapter 7   169

From (2)

$$\begin{aligned}
M'_X(\theta) &= \frac{pe^\theta}{(1-qe^\theta)^2} \\
\therefore M''_X(\theta) &= \frac{(1-qe^\theta)^2 pe^\theta - pe^\theta[-2qe^\theta(1-qe^\theta)]}{(1-qe^\theta)^4} \\
&= \frac{(1-qe^\theta)^2 pe^\theta + 2pqe^{2\theta}(1-qe^\theta)}{(1-qe^\theta)^4} \\
\therefore M''_X(0) &= \frac{(1-q)^2 p + 2pq(1-q)}{(1-q)^4} \\
&= \frac{p^3 + 2p^2 q}{p^4} = \frac{p^2(p+2q)}{p^4} = \frac{p+2q}{p^2} \\
\therefore Var(X) &= M''_X(0) - [M'_X(0)]^2 \\
&= \frac{p+2q}{p^2} - \left(\frac{1}{p}\right)^2 \\
&= \frac{p+2q-1}{p^2} = \frac{2q-(1-p)}{p^2} = \frac{2q-q}{p^2} = \frac{q}{p^2} \\
&= \frac{1-p}{p^2}
\end{aligned}$$

Moment Generating Function of Some Continuous Probability Distributions

(1) Uniform Distribution:
The probability density function of uniform distribution is given by

$$\begin{cases} \dfrac{1}{b-a} &, \ a \leq x \leq b \\ 0 &, \ \text{elsewhere} \end{cases}$$

Chapter 7

$$\text{Moment generating function } M_X(\theta) = E(e^{\theta x}) = \int e^{\theta x} f(x) dx$$

$$= \int_a^b e^{\theta x} \left(\frac{1}{b-a}\right) dx$$

$$= \frac{1}{b-a} \int_a^b e^{\theta x} dx$$

$$= \frac{1}{b-a} \left[\frac{e^{\theta x}}{\theta}\right]_a^b$$

$$= \frac{1}{(b-a)\theta} \left[e^{\theta b} - e^{\theta a}\right]$$

But

$$e^{\theta b} = 1 + \theta b + \frac{(\theta b)^2}{2!} + \frac{(\theta b)^3}{3!} + \cdots$$

$$e^{\theta a} = 1 + \theta a + \frac{(\theta a)^2}{2!} + \frac{(\theta a)^3}{3!} + \cdots$$

$$\therefore e^{\theta b} - e^{\theta a} = \theta(b-a) + \frac{\theta^2(b^2-a^2)}{2!} + \frac{\theta^3(b^3-a^3)}{3!} + \frac{\theta^4(b^4-a^4)}{4!} + \cdots$$

$$= \theta\left[(b-a) + \frac{\theta(b^2-a^2)}{2!} + \frac{\theta^2(b^3-a^3)}{3!} + \frac{\theta^3(b^4-a^4)}{4!} + \cdots\right]$$

$$\therefore M_X(\theta) = \frac{1}{(b-a)\theta} \left[e^{\theta b} - e^{\theta a}\right]$$

$$= \frac{1}{(b-a)\theta} \cdot \theta \left[(b-a) + \frac{\theta(b^2-a^2)}{2!} + \frac{\theta^2(b^3-a^3)}{3!} + \frac{\theta^3(b^4-a^4)}{4!} + \cdots\right]$$

$$\therefore M_X(\theta) = \frac{1}{b-a} \left[(b-a) + \frac{\theta(b^2-a^2)}{2!} + \frac{\theta^2(b^3-a^3)}{3!} + \frac{\theta^3(b^4-a^4)}{4!} + \cdots\right]$$

$$\therefore M_X'(\theta) = \frac{1}{b-a} \left[0 + \frac{(b^2-a^2)}{2!} + \frac{2\theta(b^3-a^3)}{3!} + \frac{3\theta^2(b^4-a^4)}{4!} + \cdots\right]$$

$$\Rightarrow M_X'(\theta) = \frac{1}{b-a} \left[\frac{b^2-a^2}{2!} + \frac{2\theta(b^3-a^3)}{3!} + \frac{3\theta^2(b^4-a^4)}{4!} + \cdots\right] \tag{1}$$

Chapter 7

From (1)

$$M''_X(\theta) = \frac{1}{b-a}\left[\frac{2(b^3-a^3)}{3!} + \frac{6\theta(b^4-a^4)}{4!} + \cdots\right]$$

$$\therefore M''_X(0) = \frac{1}{b-a}\left[\frac{b^3-a^3}{3}\right]$$

$$= \frac{1}{3(b-a)}(b^3-a^3) = \frac{1}{3(b-a)}(b-a)(b^2+ab+a^2)$$

$$= \frac{b^2+ab+a^2}{3}$$

$$\therefore Var(X) = M''_X(0) - [M'_X(0)]^2$$

$$= \frac{b^2+ab+a^2}{3} - \left[\frac{b+a}{2}\right]^2$$

$$= \frac{b^2+ab+a^2}{3} - \frac{(b^2+a^2+2ab)}{4}$$

$$= \frac{b^2-2ab+a^2}{12}$$

$$= \frac{(b-a)^2}{12}$$

(2) Exponential Distribution
The probability density function of an exponential distribution is given by

$$f(x) = \begin{cases} \lambda e^{\lambda x} &, x \geq 0 \\ 0 &, \text{otherwise} \end{cases}$$

The moment generating function for the above distribution is

$$M_X(\theta) = E(e^{\theta x}) = \int_0^\infty e^{\theta x} \lambda e^{-\lambda x} dx$$

$$= \lambda \int_0^\infty e^{-(\lambda-\theta)x} dx$$

Put $t = (\lambda - \theta)x \Rightarrow dt = (\lambda - \theta)dx$.

$$\therefore \quad dx = \frac{dt}{\lambda - \theta}$$

$$\therefore \quad M_X(\theta) = \lambda \int_0^\infty e^{-t} \frac{dt}{\lambda - \theta} = \frac{\lambda}{\lambda - \theta} \int_0^\infty e^{-t} dt = \frac{\lambda}{\lambda - \theta} \quad (1)$$

$$= \frac{\lambda}{\lambda - \theta} = \frac{1}{1 - \frac{\theta}{\lambda}} = \left(1 - \frac{\theta}{\lambda}\right)^{-1}$$

$$\therefore \quad M_X(\theta) = \left(1 - \frac{\theta}{\lambda}\right)^{-1}$$

$$M'_X(\theta) = -1\left(1 - \frac{\theta}{\lambda}\right)^{-2}\left(-\frac{1}{\lambda}\right) = \frac{1}{\lambda}\left(1 - \frac{\theta}{\lambda}\right)^{-2}$$

The mean of the distribution $= M'_X(0) = \frac{1}{\lambda}\left(1 - \frac{0}{\lambda}\right)^{-2}$

$$= \frac{1}{\lambda}(1) = \frac{1}{\lambda}$$

From (1),

$$M''_X(\theta) = -2\left(\frac{1}{\lambda}\right)\left(1 - \frac{\theta}{\lambda}\right)^{-3}\left(-\frac{1}{\lambda}\right)$$

$$= \frac{2}{\lambda^2}\left(1 - \frac{\theta}{\lambda}\right)^{-3}$$

$$M''_X(0) = \frac{2}{\lambda^2}\left(1 - \frac{0}{\lambda}\right)^{-3} = \frac{2}{\lambda^2}(1^{-3}) = \frac{2}{\lambda^2}$$

$$\therefore \quad Var(X) = M''_X(0) - [M'_X(0)]^2$$

$$= \frac{2}{\lambda^2} - \frac{1}{\lambda^2} = \frac{1}{\lambda^2}$$

Hence the variance of the distribution is $\frac{1}{\lambda^2}$.

(3) Normal Distribution

Recall that a continuous random variable X is said to have a normal distribution if and only if its probability density function is given by

$$f(x) = \frac{1}{\sigma\sqrt{2\pi}} e^{-\frac{1}{2}(\frac{x-\mu}{\sigma})^2}$$

Its moment generating function is given by

$$M_X(\theta) = e^{\mu\theta + \frac{1}{2}\sigma^2\theta^2}$$

Proof:

$$\begin{aligned}
M_X(\theta) &= \int_{-\infty}^{\infty} e^{\theta x} f(x) dx \\
&= \int_{-\infty}^{\infty} e^{\theta x} \cdot \frac{1}{\sigma\sqrt{2\pi}} e^{-\frac{1}{2}(\frac{x-u}{\sigma})^2} \\
&= \frac{1}{\sigma\sqrt{2\pi}} \int_{-\infty}^{\infty} e^{\theta x - \frac{1}{2\sigma^2}(x^2 - 2\mu x + \mu^2)} \\
&= \frac{1}{\sigma\sqrt{2\pi}} \int_{-\infty}^{\infty} \left[e^{-\frac{1}{2\sigma^2}[-2x\theta\sigma^2 + (x-\mu)^2]} \right] dx
\end{aligned}$$

Using the identity

$$-2x\theta\sigma^2 + (x-\mu)^2 = [x - (\mu + \theta\sigma^2)]^2 - 2\mu\theta\sigma^2 - \theta^2\sigma^4$$

we have

$$\begin{aligned}
M_X(\theta) &= \frac{1}{\sigma\sqrt{2\pi}} \int_{-\infty}^{\infty} e^{-\frac{1}{2\sigma^2}\{[x-(\mu+\theta\sigma^2)]^2 - (2\mu\theta\sigma^2 + \theta^2\sigma^4)\}} dx \\
&= \frac{1}{\sigma\sqrt{2\pi}} \int_{-\infty}^{\infty} e^{-\frac{1}{2}\left\{\left[\frac{x-(\mu+\theta\sigma^2)}{\sigma}\right]^2 - \frac{(2\mu\theta\sigma^2 + \theta^2\sigma^4)}{\sigma^2}\right\}} dx \\
&= \frac{1}{\sigma\sqrt{2\pi}} \int_{-\infty}^{\infty} e^{-\frac{1}{2}\left[\left(\frac{x-(\mu+\theta\sigma^2)}{\sigma}\right)^2\right] + \frac{2\mu\theta\sigma^2 + \theta^2\sigma^4}{2\sigma^2}} dx \\
&= e^{\mu\theta + \frac{1}{2}\theta^2\sigma^2} \left\{ \frac{1}{\sigma\sqrt{2\pi}} \int_{-\infty}^{\infty} e^{-\frac{1}{2}\left[\frac{x-(\mu+\theta\sigma^2)}{\sigma}\right]^2} dx \right\}
\end{aligned}$$

The expression inside the curly bracket is a probability density function of a normal distribution with parameter $\mu + \theta\sigma^2$ and σ.

Hence the moment generating function of a normal distribution is $M_X(\theta) = e^{\mu\theta + \frac{1}{2}\theta^2\sigma^2}$.

We want to use the above moment generating function to find the mean (μ) and variance (σ^2) of the given normal distribution.

From

$$\begin{aligned} M_X(\theta) &= e^{\mu\theta + \frac{1}{2}\theta^2\sigma^2} \\ M_X'(\theta) &= (\mu + \theta\sigma^2)e^{\mu\theta + \frac{1}{2}\theta^2\sigma^2} \quad (1) \\ \therefore M_X'(0) &= (\mu + 0)e^0 \\ &= \mu(1) \\ &= \mu. \end{aligned}$$

Hence the mean of the given normal distribution is μ.

From (1),

$$\begin{aligned} M_X''(\theta) &= \sigma^2 e^{\mu\theta + \frac{1}{2}\theta^2\sigma^2} + (\mu + \theta\sigma^2)(\mu + \theta\sigma^2)e^{\mu\theta + \frac{1}{2}\theta^2\sigma^2} \\ \therefore M_X''(0) &= \sigma^2 e^0 + \mu^2 e^0 \\ &= \sigma^2 + \mu^2 \end{aligned}$$

$$\begin{aligned} \text{But variance } (X) &= M_X''(0) - [M_X'(0)]^2 \\ &= \sigma^2 + \mu^2 - \mu^2 \\ &= \sigma^2 \end{aligned}$$

Hence the variance of the normal distribution is σ^2.

(4) Gamma Distribution

A random variable X has a gamma distribution if and only if its probability density function is given by

$$f(x) = \begin{cases} \dfrac{x^{\alpha-1}e^{-x/\beta}}{\Gamma(\alpha)\beta^\alpha} & , x > 0 \\ 0 & , \text{elsewhere.} \end{cases}$$

Its moment generating function is given by

$$\begin{aligned}
M_X(\theta) &= \int e^{\theta x} f(x)dx \\
&= \int_0^\infty e^{\theta x} \left(\frac{1}{\Gamma(\alpha)\beta^\alpha} x^{\alpha-1} e^{-x/\beta}\right) dx \\
&= \frac{1}{\Gamma(\alpha)\beta^\alpha} \int_0^\infty e^{\theta x} x^{\alpha-1} e^{-x/\beta} dx \\
&= \frac{1}{\Gamma(\alpha)\beta^\alpha} \int_0^\infty x^{\alpha-1} e^{-x(\frac{1}{\beta}-\theta)} dx \\
&= \frac{1}{\Gamma(\alpha)\beta^\alpha} \int_0^\infty x^{\alpha-1} e^{-\frac{x}{\beta}(1-\theta\beta)} dx
\end{aligned}$$

Let $u = \dfrac{x}{\beta}(1-\theta\beta)$.

$$\therefore\ x = \frac{\beta u}{1-\theta\beta}\ \text{ and }\ dx = \frac{\beta du}{1-\theta\beta}.$$

$$\begin{aligned}
\therefore\ M_X(\theta) &= \frac{1}{\Gamma(\alpha)\beta^\alpha} \int_0^\infty \left(\frac{\beta u}{1-\beta\theta}\right)^{\alpha-1} e^{-u} \frac{\beta du}{1-\beta\theta} \\
&= \frac{1}{\Gamma(\alpha)\beta^\alpha} \int_0^\infty \frac{(\beta^{\alpha-1} u^{\alpha-1} \cdot \beta)}{(1-\beta\theta)^{\alpha-1}(1-\beta\theta)} e^{-u} du \\
&= \frac{1}{\Gamma(\alpha)\beta^\alpha} \int_0^\infty \frac{\beta^\alpha u^{\alpha-1}}{(1-\beta\theta)^\alpha} e^{-u} du \\
&= \frac{1}{\Gamma(\alpha)\beta^\alpha} \left(\frac{\beta^\alpha}{(1-\beta\theta)^\alpha}\right) \int_0^\infty u^{\alpha-1} e^{-u} du \\
&= \frac{1}{\Gamma(\alpha)(1-\beta\theta)^\alpha}\Gamma(\alpha) = \frac{1}{\Gamma(\alpha)(1-\beta\theta)^\alpha}\Gamma(\alpha) \\
&= \frac{1}{(1-\beta\theta)^\alpha} = (1-\beta\theta)^{-\alpha}
\end{aligned}$$

$$\therefore\ M_X(\theta) = (1-\beta\theta)^{-\alpha}$$

We want to find the mean and variance of the distribution from $M_X(\theta)$.

$$\begin{aligned} M'_X(\theta) &= -\alpha(-\beta)(1-\beta\theta)^{-(\alpha+1)} = \alpha\beta(1-\beta\theta)^{-(\alpha+1)} \\ M'_X(0) &= \alpha\beta(1)^{-(\alpha+1)} = \alpha\beta \end{aligned} \qquad (1)$$

Hence mean of gamma distribution $= \alpha\beta$.
From (1),

$$\begin{aligned} M''_X(\theta) &= \alpha\beta[-(\alpha+1)(-\beta)](1-\theta\beta)^{-(\alpha+2)} \\ &= \alpha\beta^2(\alpha+1)(1-\theta\beta)^{-(\alpha+2)} \\ &= \alpha(\alpha+1)\beta^2(1-\theta\beta)^{-(\alpha+2)} \\ M''_X(0) &= \alpha(\alpha+1)\beta^2(1)^{-(\alpha+2)} \\ &= \alpha(\alpha+1)\beta^2. \\ \therefore Var(X) &= M''_X(0) - [M'_X(0)]^2 \\ &= \alpha(\alpha+1)\beta^2 - (\alpha\beta)^2 \\ &= \alpha^2\beta^2 + \alpha\beta^2 - \alpha^2\beta^2 \\ &= \alpha\beta^2. \end{aligned}$$

Hence variance of a random variable with gamma distribution is $\alpha\beta^2$.

(5) Chi-square Distribution
A random variable X is said to have a chi-square distribution if and only if its probability density function is given by

$$f(x) = \begin{cases} \dfrac{x^{\frac{r}{2}-1}e^{-x/2}}{\Gamma\left(\frac{r}{2}\right)2^{\frac{r}{2}}} & , x > 0 \\ 0 & , \text{elsewhere} \end{cases}$$

Chi-square distribution is a special case of Gamma distribution in which $\alpha = \dfrac{r}{2}$ and $\beta = 2$.
With the above relationship, it becomes clear that we can obtain moment generating function of chi-square from Gamma distribution.
Since the moment generating function of a Gamma distribution

$M_X(\theta) = (1 - \beta\theta)^{-\alpha}$.

∴ substituting $\frac{r}{2}$ for α and 2 for β, the moment generating function of chi-square distribution becomes:

$$\begin{aligned} M_X(\theta) &= (1-2\theta)^{-r/2} \\ \therefore M'_X(\theta) &= -\frac{r}{2}(-2)(1-2\theta)^{-(\frac{r}{2}+1)} = r(1-2\theta)^{-(\frac{r}{2}+1)} \qquad (1) \\ M'_X(0) &= r(1)^{-(\frac{r}{2}+1)} = r = E(X) \end{aligned}$$

∴ The mean of a chi-square distribution $\mu = r$.
From (1),

$$\begin{aligned} M''_X(\theta) &= r\left(-(\frac{r}{2}+1)(-2)(1-2\theta)^{-(\frac{r}{2}+2)}\right) \\ &= 2r(\frac{r}{2}+1)(1-2\theta)^{-(\frac{r}{2}+2)} \\ M''_X(0) &= 2r(\frac{r}{2}+1) \\ \therefore Var(X) &= M''_X(0) - [M'_X(0)]^2 \\ &= 2r(\frac{r}{2}+1) - r^2 \\ &= r^2 + 2r - r^2 \\ &= 2r. \end{aligned}$$

∴ Variance of a chi-square distribution is $\sigma^2 = 2r$.

Theorem
The moment generating function of the sum of two independent random variables is the product of their moment generating functions.

Solution
If X, Y are independent, we want to show that the m.g.f. of $W = X + Y$ is
$$M_W(\theta) = M_X(\theta) \cdot M_Y(\theta)$$
where $M_X(\theta)$ is the m.g.f. of X and $M_Y(\theta)$ is the m.g.f. of Y.

Proof
(1) For continuous random variable X and Y
Joint pdf of X, Y is $f(x,y)$

$$\begin{aligned}
\therefore \quad M_W(\theta) &= E(e^{\theta w}) \\
&= \int_y \int_x e^{\theta(x+y)} f(x,y) dx dy \\
&= \int_y^\infty \int_x^\infty e^{\theta x} e^{\theta y} f(x,y) dx dy \\
&= \int_y \int_x e^{\theta x} e^{\theta y} f_x(x) f(y) dx dy \quad (X, Y \text{ are independent}) \\
&= \int_x e^{\theta x} f_x(x) dx \int_y e^{\theta y} f_y(y) dy \\
&= M_X(\theta) \cdot M_Y(\theta) \quad \text{proved.}
\end{aligned}$$

(ii) For discrete random variables

$$\begin{aligned}
M_W(\theta) &= \sum_x \sum_y e^{\theta(x+y)} f(x,y) \\
&= \sum_x \sum_y e^{\theta x + \theta y} f(x,y) \\
&= \sum_x \sum_y e^{\theta x} e^{\theta y} f(x,y) \\
&= \sum_x \sum_y e^{\theta x} e^{\theta y} f(x) f(y) \quad (X, Y \text{ are independent}) \\
&= \sum_x e^{\theta x} f(x) \sum_y e^{\theta y} f(y) \\
&= E(e^{\theta x}) \cdot E(e^{\theta y}) \\
&= M_X(\theta) \cdot M_y(\theta) \quad \text{proved}
\end{aligned}$$

Proofs based on the above theorem
Required:- To show that

(i) $M_{X+a}(\theta) = e^{\theta a} M_X(\theta)$

Chapter 7

(ii) $M_{\frac{X+a}{b}}(\theta) = e^{\frac{\theta a}{b}} M_X(\theta/b)$

(iii) $M_{bX}(\theta) = E(e^{bx\theta}) = M_X(b\theta)$

Proof:

(i) For a discrete random variable X

$$\begin{aligned}
M_{X+a}(\theta) &= E[e^{\theta(x+a)}] \\
&= \sum e^{\theta(x+a)} f(x) \\
&= \sum e^{\theta x} \cdot e^{\theta a} f(x) \\
&= e^{\theta a} \sum e^{\theta x} f(x) \\
&= e^{\theta a} E(e^{\theta x}) \\
&= e^{\theta a} M_X(\theta) \quad \text{proved.}
\end{aligned}$$

For a continuous random variable X

$$\begin{aligned}
M_{X+a}(\theta) &= E[e^{\theta(x+a)}] \\
&= \int e^{\theta(x+a)} f(x) dx \\
&= \int e^{\theta x + \theta a} f(x) dx \\
&= \int e^{\theta x} \cdot e^{\theta a} f(x) dx \\
&= e^{\theta a} \int e^{\theta x} f(x) dx \\
&= e^{\theta a} E(e^{\theta x}) \\
&= e^{\theta a} M_X(\theta) \quad \text{proved.}
\end{aligned}$$

(ii) For a discrete random variable

$$\begin{aligned}
M_{\frac{X+a}{b}}(\theta) &= E\left[e^{\theta\left(\frac{x+a}{b}\right)}\right] \\
&= \sum e^{\theta\left(\frac{x+a}{b}\right)} f(x) \\
&= \sum e^{\frac{\theta x + \theta a}{b}} f(x) \\
&= \sum e^{\frac{\theta x}{b} + \frac{\theta a}{b}} f(x) \\
&= \sum e^{\frac{\theta x}{b}} \cdot e^{\frac{\theta a}{b}} f(x) \\
&= e^{\frac{\theta a}{b}} \sum e^{\frac{\theta x}{b}} f(x) \\
&= e^{\frac{\theta a}{b}} E\left(e^{\frac{\theta x}{b}}\right) \\
&= e^{\frac{\theta a}{b}} M_x(\theta/b) \quad \text{proved}
\end{aligned}$$

For a continuous random variable,

$$\begin{aligned}
M_{\frac{X+a}{b}}(\theta) &= E\left[e^{\frac{\theta(x+a)}{b}}\right] \\
&= \int e^{\frac{\theta(x+a)}{b}} f(x)\, dx \\
&= \int e^{\frac{\theta x + \theta a}{b}} f(x)\, dx \\
&= \int e^{\frac{\theta x}{b} + \frac{\theta a}{b}} f(x)\, dx \\
&= \int \left(e^{\frac{\theta x}{b}} \cdot e^{\frac{\theta a}{b}}\right) f(x) dx \\
&= e^{\frac{\theta a}{b}} \int e^{\frac{\theta x}{b}} f(x) dx \\
&= e^{\frac{\theta a}{b}} \cdot E\left(e^{\frac{\theta x}{b}}\right) \\
&= e^{\frac{\theta a}{b}} M_X\left(\frac{\theta}{b}\right) \quad \text{proved.}
\end{aligned}$$

(iii) For a discrete random variable

$$\begin{aligned} M_{bX}(\theta) = E(e^{bx\theta}) &= \sum e^{bx\theta} f(x) \\ &= \sum e^{b\theta x} f(x) \\ &= E(e^{b\theta x}) \\ &= M_X(b\theta) \quad \text{proved.} \end{aligned}$$

For a continuous random variable

$$\begin{aligned} M_{bx}(\theta) &= E(e^{bx\theta}) \\ &= \int e^{bx\theta} f(x) dx \\ &= \int e^{b\theta x} f(x) dx \\ &= E(e^{b\theta x}) \\ &= M_X(b\theta) \quad \text{proved.} \end{aligned}$$

Cummulant Generating Function

This is defined as the natural logarithm of moment generating function. Hence if $C(t)$ and $M(t)$ are cummulant generating function and moment generating function respectively, $C(t) = \ln M(t)$. In other words

$$M(t) = e^{C(t)}.$$

From $C(t) = \ln M(t)$

$$\frac{dC(t)}{dt} = \frac{M'(t)}{M(t)} \tag{1}$$

$$\frac{dC(t)}{dt}_{(t=0)} = \frac{M'(0)}{M(0)}$$

Since $M(t) = E(e^{tx})$, $M(0) = E(e^0) = 1$.

$$\therefore \quad \frac{dC(t)}{dt}_{(t=0)} = \frac{M'(0)}{1} = M'(0).$$

By definition, $M'(0) = \mu$,
$$\therefore \frac{dC(t)}{dt}_{(t=0)} = \mu.$$

From $\dfrac{dC(t)}{dt} = \dfrac{M'(t)}{M(t)}$,

$$\frac{d^2C(t)}{dt^2} = \frac{M(t)M''(t) - M'(t) \cdot M'(t)}{[M(t)]^2}$$
$$= \frac{M(t)M''(t) - [M'(t)]^2}{[M(t)]^2}$$
$$\therefore \frac{d^2C(t)}{dt^2_{(t=0)}} = \frac{M(0)M''(0) - [M'(0)]^2}{[M(0)]^2}$$

Since $M(0) = 1$,
$$\therefore \frac{d^2C(t)}{dt^2_{(t=0)}} = M''(0) - [M'(0)]^2 = \sigma^2$$
$$\therefore \frac{d^2C(t)}{dt^2_{(t=0)}} = \sigma^2.$$

Hence the first two derivatives of cummulant generating function evaluated at $t = 0$ gives the mean and variance of the distribution respectively, i.e. the first derivative at $t = 0$ gives the mean of the distribution and the second derivative at $t = 0$ gives the variance of the distribution.

Example 1.
A random variable X has a moment generating function
$$M(t) = \exp[2(e^t - 1)].$$

Find the cummulant generating function of X and use it to evaluate the mean and variance of X.

Solution: By definition,
$$C(t) = \ln M(t)$$

Chapter 7

Since $M(t) = \exp[2(e^t - 1)]$

$$\begin{aligned}\therefore\quad C(t) &= \ln \exp[2(e^t - 1)] \\ &= \ln e^{2(e^t-1)} \\ &= 2(e^t - 1)\end{aligned}$$

$$\therefore\quad \frac{dC(t)}{dt} = 2e^t$$

$$\therefore\quad \text{Mean} = \frac{dC(t)}{dt}_{(t=0)} = 2e^0 = 2.$$

$$\frac{d^2C(t)}{dt^2} = 2e^t$$

$$\therefore\quad \frac{d^2C(t)}{dt^2}_{(t=0)} = 2e^0 = 2 = \sigma^2$$

Hence the mean of $X = 2$ and its variance $= 2$.

Example 2.
Find the cummulant generating function of a probability distribution whose moment generating function $(mgf) = (pe^t + q)^n$ where $p + q = 1$. Identify the distribution.

Solution:

$$\begin{aligned}M(t) &= (pe^t + q)^n \\ \therefore\quad C(t) &= \ln M(t) = \ln(pe^t + q)^n.\end{aligned}$$

$$\therefore\quad \frac{dC(t)}{dt} = \frac{npe^t(pe^t + q)^{n-1}}{(pe^t + q)^n} \qquad (1)$$

$$\begin{aligned}\text{The mean of the distribution} &= \frac{dC(t)}{dt(t=0)} \\ &= \frac{npe^0[pe^0 + q]^{n-1}}{(pe^0 + q)^n} = \frac{np(p+q)^{n-1}}{(p+q)^n} \\ &= \frac{np}{1} = np\end{aligned}$$

∴ The mean of the distribution is np.

From $\dfrac{dC(t)}{dt} = \dfrac{npe^t(pe^t+q)^{n-1}}{(pe^t+q)^n}$

$\dfrac{d^2C(t)}{dt^2} = \dfrac{(pe^t+q)^n[npe^t(pe^t+q)^{n-1} + npe^t[(n-1)pe^t(pe^t+q)^{n-2}]}{[(pe^t+q)^n]^2}$
$\phantom{\dfrac{d^2C(t)}{dt^2}} \quad \dfrac{-[npe^t(pe^t+q)^{n-1}npe^t(pe^t+q)^{n-1}]}{[(pe^t+q)^n]^2}$

$\dfrac{d^2C(t)}{dt^2}\bigg|_{t=0} = \dfrac{(p+q)^n[np(p+q)^{n-1} + n(n-1)p^2(p+q)^{n-2}}{(p+q)^{2n}}$
$\phantom{\dfrac{d^2C(t)}{dt^2}\bigg|_{t=0}} \quad \dfrac{-np(p+q)^{n-1}np(p+q)^{n-1}]}{(p+q)^{2n}}$

$\phantom{\dfrac{d^2C(t)}{dt^2}\bigg|_{t=0}} = np + n(n-1)p^2 - n^2p^2$
$\phantom{\dfrac{d^2C(t)}{dt^2}\bigg|_{t=0}} = np + n^2p^2 - np^2 - n^2p^2$
$\phantom{\dfrac{d^2C(t)}{dt^2}\bigg|_{t=0}} = np - np^2$
$\phantom{\dfrac{d^2C(t)}{dt^2}\bigg|_{t=0}} = np(1-p)$
$\phantom{\dfrac{d^2C(t)}{dt^2}\bigg|_{t=0}} = npq.$

Hence, the mean and variance of the distribution are respectively np and npq. The probability distribution whose mean is np and variance is npq is Binomial distribution.

Example 3.
Find the cummulant generating function of a Poisson distribution and use it to find the mean and the variance of the distribution.

Solution:
The moment generating function $M(t)$ of a Poisson distribution $= e^{\lambda(e^t-1)}$.

Since $M(t) = e^{\lambda(e^t-1)}$, cummulant generating function

$$\begin{aligned} C(t) &= \ln M(t) \\ &= \ln e^{\lambda(e^t-1)} \\ &= \lambda(e^t - 1) = \lambda e^t - \lambda \end{aligned}$$

$$\therefore \frac{dC(t)}{dt} = \lambda e^t$$

$$\therefore \frac{dC(t)}{dt}_{(t=0)} = \lambda e^0 = \lambda.$$

Hence the mean of Poisson distribution is λ.
From

$$\frac{dC(t)}{dt} = \lambda e^t$$

$$\frac{d^2C(t)}{dt^2} = \lambda e^t$$

$$\therefore \frac{d^2C(t)}{dt^2}_{(t=0)} = \lambda e^0 = \lambda.$$

Hence the variance of Poisson distribution is λ.

Example 4.
From the moment generating function of an exponential distribution, find the cummulant generating function. Hence find its mean and variance.

Solution:
For exponential distribution with parameter λ, the moment generating function $M(t) = \left(1 - \frac{t}{\lambda}\right)^{-1}$.

∴ The cummulant generating function

$$C(t) = \ln M(t)$$
$$= \ln\left(1 - \frac{t}{\lambda}\right)^{-1}$$
$$\therefore C(t) = \ln\left(1 - \frac{t}{\lambda}\right)^{-1} \qquad (1)$$

From (1), $\dfrac{dC(t)}{dt} = \dfrac{-1\left(-\frac{1}{\lambda}\right)\left(1 - \frac{t}{\lambda}\right)^{-2}}{\left(1 - \frac{t}{\lambda}\right)^{-1}} = \dfrac{1}{\lambda}\left(1 - \frac{t}{\lambda}\right)^{-1}$

$$= \frac{1}{\lambda}\left(1 - \frac{t}{\lambda}\right)^{-1} \qquad (2)$$

$\therefore \dfrac{dC(t)}{dt(t=0)} = \dfrac{1}{\lambda}(1) = \dfrac{1}{\lambda}$

Hence the mean of the given exponential distribution is $\dfrac{1}{\lambda}$.

From (2), $\dfrac{d^2C(t)}{dt^2} = -\dfrac{1}{\lambda}\left(-\dfrac{1}{\lambda}\right)\left(1 - \dfrac{t}{\lambda}\right)^{-2} = \dfrac{1}{\lambda^2}\left(1 - \dfrac{t}{\lambda}\right)^{-2}$

$\therefore \dfrac{d^2C(t)}{dt^2_{(t=0)}} = \dfrac{1}{\lambda^2}(1 - 0)^{-2} = \dfrac{1}{\lambda^2}.$

Hence the variance of the given exponential distribution is $\dfrac{1}{\lambda^2}$.

Example 5:
Use the cummulant generating function of a Gamma distribution to show that its mean is $\dfrac{n}{\lambda}$ and its variance is $\dfrac{n}{\lambda^2}$.

Solution:
The moment generating function of a Gamma distribution
$M(t) = \left(1 - \dfrac{t}{\lambda}\right)^{-n}.$

Chapter 7

∴ The cummulant generating function

$$C(t) = \ln M(t) = \ln\left(1 - \frac{t}{\lambda}\right)^{-n} \tag{1}$$

From (1), $\dfrac{dC(t)}{dt} = \dfrac{\dfrac{n}{\lambda}\left(1 - \dfrac{t}{\lambda}\right)^{-(n+1)}}{\left(1 - \dfrac{t}{\lambda}\right)^{-n}} = \dfrac{\dfrac{n}{\lambda}}{1 - \dfrac{t}{\lambda}}$ (2)

$$\therefore \frac{dC(t)}{dt}_{(t=0)} = \frac{\frac{n}{\lambda}}{1-0} = \frac{n}{\lambda}.$$

Hence the mean of gamma distribution is $\dfrac{n}{\lambda}$.

From (2),

$$\frac{dC(t)}{dt} = \frac{\frac{n}{\lambda}}{1 - t/\lambda} = \frac{n}{\lambda}\left(1 - \frac{t}{\lambda}\right)^{-1}$$

$$\therefore \frac{d^2C(t)}{dt^2} = -\frac{n}{\lambda}\left(-\frac{1}{\lambda}\right)\left(1 - \frac{t}{\lambda}\right)^{-2}$$

$$\therefore \frac{d^2C(t)}{dt^2} = \frac{n}{\lambda^2}\left(1 - \frac{t}{\lambda}\right)^{-2}$$

$$\therefore \frac{d^2C(t)}{dt^2_{t=0}} = \frac{n}{\lambda^2}(1)^{-2} = \frac{n}{\lambda^2}.$$

Hence the variance of gamma distribution is $\dfrac{n}{\lambda^2}$.

Example 6.
Find the cummulant generating function of a uniform distribution which lies in the interval $a < x < b$. Use your result to find the mean of the distribution.
Solution:
The probability density function of a uniform distribution which lies in the

Chapter 7

interval $a < x < b$ is given by

$$f(x) = \begin{cases} \dfrac{1}{b-a} & , a < x < b \\ 0 & , \text{elsewhere} \end{cases}$$

Its moment generating function is by

$$M(t) = \frac{1}{t(b-a)}\left[e^{bt} - e^{at}\right]$$

$$= \frac{1}{b-a}\left[b - a + \frac{t(b^2 - a^2)}{2!} + \frac{t^2(b^3 - a^3)}{3!} + \cdots + \frac{t^{r-1}(b^r - a^r)}{r!} + \cdots\right]$$

$$\therefore\ C(t) = \ln M(t)$$

$$= \ln \frac{1}{b-a}\left(b - a + \frac{t(b^2 - a^2)}{2!} + \frac{t^2(b^3 - a^3)}{3!} + \cdots + \frac{t^{r-1}(b^r - a^r)}{r!} + \cdots\right]$$

$$\therefore\ \frac{dC(t)}{dt} = \frac{\dfrac{1}{b-a}\left[\dfrac{b^2 - a^2}{2!} + \dfrac{2t(b^3 - a^3)}{3!} + \cdots + \dfrac{(r-1)t^{r-2}(b^r - a^r)}{r!} + \cdots\right]}{\dfrac{1}{b-a}\left[b - a + \dfrac{t(b^2 - a^2)}{2!} + \dfrac{t^2(b^3 - a^3)}{3!} + \cdots + \dfrac{t^{r-1}(b^r - a^r)}{r!} + \cdots\right]}$$

$$\therefore\ \frac{dC(t)}{dt(t=0)} = \frac{\dfrac{b^2 - a^2}{2!}}{b - a} = \frac{b^2 - a^2}{2(b-a)} = \frac{1}{2}(b+a).$$

Hence the mean of the given uniform distribution is $\dfrac{b+a}{2}$.

The Characteristics Functions
Definition
  The characteristic function (C.F.) is defined as

$$\phi(t) = E(e^{itx}) \qquad \text{where } t = \sqrt{-1}$$

$$\phi(t) = E\left\{\sum_{r=0}^{\infty} \frac{(it)^r x^r}{r!}\right\}$$

$$= \sum_{r=0}^{\infty} \frac{(it)^r}{r!}\mu'_r$$

Chapter 7

⇒ In the power series expansion of $\phi(t)$, μ'_r is the coefficient of $\dfrac{(it)^r}{r!}$ and the rth moment about the origin.

Properties of Characteristic Functions

(i) For each p.d.f., $\exists$ a unique Characteristic Function $(\phi(t))$ and for each $\phi(t)$ $\exists$ a unique p.d.f. $[f(x)]$.

(ii) If $\phi_x(t)$, $\phi_y(t)$ are characteristic functions of X, Y respectively and if X, Y are independent, then Characteristic Function $\phi(t)$ of $(X + Y)$ is given by
$$\phi(t) = \phi_z(t)\phi_y(t).$$

(iii) Let $F_1(x), F_2(x), F_3(x)$, be a sequence of cumulative distribution functions.
Let $\phi_1(t), \phi_2(t), \phi_3(t), \cdots$ be the corresponding sequence of C.F.'s, if there exists $\phi(t)$ such that $\phi(t) = \lim\limits_{n \to \infty} \phi_n(t)$, then there must exist distribution function $F(x)$ such that
$$F(x) = \lim_{n \to \infty} F_n(x)$$

Example
Given that $f(x) = \lambda e^{-\lambda x}$, find its characteristic function

$$\begin{aligned}
f(x) &= \lambda e^{-\lambda x} \\
M(it) &= E(e^{itx}) \\
&= \int e^{itx} f(x) dx \\
&= \int e^{itx} \lambda e^{-\lambda x} dx \\
&= \lambda \int e^{-(\lambda - it)} dx
\end{aligned}$$

Chapter 7

Put $u = (\lambda - it)x \quad \therefore \quad du = (\lambda - it)dx$

$$\begin{aligned} M(it) &= \lambda \int_0^\infty e^{-u} \frac{du}{\lambda - it} \\ &= \frac{\lambda}{\lambda - it} \int_0^\infty e^{-u} du \\ &= \frac{\lambda}{\lambda - it} \Gamma(t) = \frac{\lambda}{\lambda - it} = \frac{1}{1 - \frac{it}{\lambda}} = \left(1 - \frac{it}{\lambda}\right)^{-1} \end{aligned}$$

The required characteristic function is $\left(1 - \dfrac{it}{\lambda}\right)^{-1}$.

Similar steps can be taken for other distributions.

Chapter 8

Bivariate Normal Distribution and Stochastic Independence

Definition

The continuous random variables X and Y are said to have a bivariate normal distribution if their joint density function is given by:

$$f(x,y) = \frac{1}{2\pi\sigma_x\sigma_y\sqrt{1-\rho^2}} \exp\left\{\frac{-1}{2(1-\rho^2)}\left[\left(\frac{x-\mu_x}{\sigma_x}\right)^2 - 2\rho\left(\frac{x-\mu_x}{\sigma_x}\right)\left(\frac{y-\mu_y}{\sigma_y}\right) + \left(\frac{y-\mu_y}{\sigma_y}\right)^2\right]\right\}$$

where $-\infty < x < \infty$, $-\infty < y < \infty$; μ_x, μ_y are the means of X and Y respectively. σ_x, σ_y are the corresponding standard deviations of X and Y and ρ is the correlation coefficient between X and Y.

Points to Note on Bivariate Normal Distribution

(1) If (X, Y) is bivariate normal with $\rho_{x,y} = 0$, then X and Y are independent

(2) The marginal distributions of a bivariate normal distribution are normal.

(3) If (U, V) is obtained from the bivariate normal (X, Y) by linear transformation, then (U, V) is bivariate normal.

Chapter 8

(4) The continuous random variables X and Y are said to have a bivariate normal distribution if and only if the combination $aX + bY$ is normal for every a and b.

(5) The characteristic function of a bivariate normal distribution can be derived from the characteristic function of the univariate normal distribution. Thus if (X, Y) is bivariate normal, then for any a and b

$$\Phi(ax+by)(t) = E[e^{t(aX+bY)}] = \Phi_{X,Y}(at, bt),$$
$$= \exp\left(it\mu - \frac{1}{2}\sigma^2 t^2\right)$$

where
$$\mu = a\mu_x X + b\mu_y Y$$

and
$$\sigma^2 = a^2\sigma_x^2 + b^2\sigma_y^2 + 2ab\sigma_{xy}.$$

The level curves of a bivariate normal density for (X, Y) are ellipses centred at (μ_x, μ_y). The axes are parallel to the coordinate axes if and only if X and Y are independent. With a proper choice of multiplicative constant, any function of the form

$$f(x, y) = \text{const } \exp\left[-\frac{1}{2}(ax^2 + 2bxy + cy^2)\right]$$

can serve as the density of a bivariate normal distribution if $a > 0$ and $ac > b^2$ (so that the level curves are ellipses). It is necessary to identify the coefficients in the quadratic density. Doing so yields

$$\sigma_x^2 = \frac{a}{ac - b^2}, \quad \sigma_y^2 = \frac{a}{ac - b^2}, \quad \rho = \frac{-b}{\sqrt{ab}}.$$

The Moment Generating Function of the Bivariate Normal Distribution

Recall that if X and Y are independent random variables, the moment generating function of the sum $X + Y$ is particularly simple combination of the

moment generating functions of the summands, namely their product.
If $\varphi_{X+Y}(t)$ is the moment generating function of the sum $X + Y$, then

$$\begin{aligned}\varphi_{X+Y}(t) &= E[e^{t(x+y)}] = E(e^{tx}e^{ty}) \\ &= E(e^{tx})E(e^{ty}) = \varphi_x(t)\varphi_y(t)\end{aligned}$$

If (X, Y) is bivariate normal distribution, the moment generating function of X is given by

$$M_X(t) = e^{\mu t}e^{\sigma^2 \frac{t^2}{2}}\frac{1}{\sqrt{2\pi}\sigma}\int_{-\infty}^{\infty}e^{\frac{(x-\mu-\sigma^2 t)^2}{2\sigma^2}}dx$$

and m.g.f. of Y is given by

$$M_Y(t) = e^{\mu t}e^{\sigma^2 \frac{t^2}{2}}\frac{1}{\sqrt{2\pi}\sigma}\int_{-\infty}^{\infty}e^{\frac{(y-\mu-\sigma^2 t)^2}{2\sigma^2}}dy$$

The m.g.f. of the sum $X + Y$ is given by $M_{(X+Y)}(t) = M_X(t); M_Y(t)$ where $M_x(t)$ and $M_y(t)$ are as defined above.

$$\begin{aligned}M_{(X+Y)}(t) &= M_X(t) \cdot M_Y(t) \\ &= e^{\mu t}e^{\sigma^2 \frac{t^2}{2}}\frac{1}{\sqrt{2\pi}\sigma}\int_{-\infty}^{\infty}e^{\frac{(x-\mu-\sigma^2 t)^2}{2\sigma^2}}dx \cdot e^{\mu t}e^{\sigma^2 \frac{t^2}{2}} \cdot \frac{1}{\sqrt{2\pi}\sigma}\int_{-\infty}^{\infty}e^{\frac{(y-\mu-\sigma^2 t)^2}{2\sigma^2}}dy \\ &= e^{\mu t}e^{\sigma^2 \frac{t^2}{2}} \cdot e^{\mu t} \cdot e^{\sigma^2 \frac{t^2}{2}} \cdot \frac{1}{\sqrt{2\pi}\sigma} \cdot \frac{1}{\sqrt{2\pi}\sigma}\int_{-\infty}^{\infty}\int_{-\infty}^{\infty}e^{\frac{(x-\mu-\sigma^2 t)^2}{2\sigma^2}} \cdot e^{\frac{(y-\mu-\sigma^2 t)^2}{2\sigma^2}}dxdy \\ &= e^{2\mu t+2\sigma^2 \frac{t^2}{2}}\frac{1}{2\pi\sigma}\int_{-\infty}^{\infty}\int_{-\infty}^{\infty}e^{\frac{1}{2\sigma^2}[(x-\mu-\sigma^2 t)^2+(y-\mu-\sigma^2 t)^2]}dxdy \\ &= \frac{1}{2\pi\sigma}e^{2\mu t+2\sigma^2 \frac{t^2}{2}}\int_{-\infty}^{\infty}\int_{-\infty}^{\infty}e^{\frac{1}{2\sigma^2}[(x-\mu-\sigma^2 t)^2+(y-\mu-\sigma^2 t)^2]}dxdy\end{aligned}$$

From the above $X + Y \sim N(2\mu; 2\sigma^2)$.

Stochastic Independence

Let X_1 and X_2 denote random variables having the joint probability density function $f(x_1, x_2)$ and the marginal probability density functions $f_1(x_1)$ and

$f_2(x_2)$ respectively. The random variables X_1 and X_2 are said to be stochastically independent if and only if $f(x_1, x_2) = f_1(x_1) \cdot f_2(x_2)$.

Theorem: If X_1 and X_2 are stochastically independent random variables with marginal probability density functions $f_1(x_1)$ and $f_2(x_2)$ respectively, then

$$P[a < x_1 < b, \ c < x_2 < d] = P[a < x_1 < b] \cdot P[c < x_2 < d]$$

for every $a < b$ and $c < d$ where a, b, c and d constants.

Proof: Since X_1 and X_2 are stochastically independent, we have $f(x_1, x_2) = f_1(x_1) \cdot f_2(x_2)$.

$$\begin{aligned} P[a < x_1 < b, \ c < x_2 < d] &= \int_a^b \int_c^d f(x_1, x_2) dx_1 dx_2 \\ &= \left[\int_a^b f_1(x_1) dx_1\right] \left[\int_c^d f_2(x_2) dx_2\right] \\ &= P(a < X_1 < b) P(c < X_2 < d) \quad \text{proved.} \end{aligned}$$

The Distribution of Sum of Two or More Independent Random Variables

For continuous Random Variables:

Theorem: Suppose X and Y are independent continuous random variables with probability density functions g and h respectively where $Z = X + Y$ and $W = X$

$$q(Z) = \int_{-\infty}^{\infty} g(w) h(z - w) dw.$$

The above is convolution theorem for sum of two independent continuous random variables.

Proof: Since X and Y are independent the joint probability density function may be written as $f(x, y) = g(x) h(y)$.

Now, consider the transformation $z = x + y$, $w = x$ then $x = w$ and $y = z - w$.

The Jacobian of the transformation is

$$J(z,w) = \frac{\partial(x,y)}{\partial(z,w)} = \begin{vmatrix} \frac{\partial x}{\partial z} & \frac{\partial x}{\partial w} \\ \frac{\partial y}{\partial z} & \frac{\partial y}{\partial w} \end{vmatrix}$$

$$= \begin{vmatrix} 0 & 1 \\ 1 & -1 \end{vmatrix} = |-1| = 1.$$

$$\therefore \quad |J(z,w)| = |-1| = 1.$$

Hence the joint probability density function of $Z = X+Y$ and $w = x$ is

$$K(z,w) = g(w)h(z-w)|J(z,w)|$$

$$\therefore \quad \int_{-\infty}^{\infty} K(z,w)dw = \int_{-\infty}^{\infty} g(w)h(z-w)dw$$

Put $q(z) = \int_{-\infty}^{\infty} K(z,w)dw$, $\therefore \; q(z) = \int_{-\infty}^{\infty} g(w)h(z-w)dw$ proved.

Example 1

Let X and Y be independent random variables each with distribution $N(0,1)$. Find the probability density function of $Z = X+Y$.
Show that the probability density function of $Z = X+Y$ is $Z \sim N(0,2)$.

Solution

Since the random variables are from $N(0,1)$,

$$g(x) = \frac{1}{\sqrt{2\pi}} e^{-x^2/2} \quad -\infty < x < \infty$$

and

$$h(y) = \frac{1}{\sqrt{2\pi}} e^{-y^2/2} \quad -\infty < y < \infty.$$

Chapter 8

∴ If $w = x, \Rightarrow y = z - w$ then

$$\begin{aligned}
f(z) &= \int_{-\infty}^{\infty} g(w)h(z-w)dw. \\
\therefore f(z) &= \int_{-\infty}^{\infty} \frac{1}{\sqrt{2\pi}}e^{-w^2/2}\frac{1}{\sqrt{2\pi}}e^{-(z-w)^2/2}dw \\
&= \frac{1}{2\pi}\int_{-\infty}^{\infty} e^{-\frac{1}{2}[w^2+(z-w)^2]}dw \\
&= \frac{1}{2\pi}\int_{-\infty}^{\infty} e^{-\frac{1}{2}(w^2+z^2+w^2-2wz)}dw \\
&= \frac{1}{2\pi}\int_{-\infty}^{\infty} e^{-\frac{1}{2}(2w^2+z^2-2wz)}dw \\
&= \frac{1}{2\pi}e^{-z^2/2}\int_{-\infty}^{\infty} e^{-(w^2-wz)}dw \\
&= \frac{e^{-z^2/2}}{2\pi}e^{z^2/4}\int_{-\infty}^{\infty} e^{-(w-\frac{z}{2})^2}dw \\
&= \frac{e^{-z^2/4}}{2\pi}\int_{-\infty}^{\infty} e^{-\frac{1}{2}[2(w-\frac{z}{2})^2]}dw \\
&= \frac{e^{-z^2/4}}{2\pi}\int_{-\infty}^{\infty} e^{-\frac{1}{2}[\sqrt{2}(w-\frac{z}{2})]^2}dw
\end{aligned}$$

Set $t = \sqrt{2}(w - \frac{z}{2}) \Rightarrow dt = \sqrt{2}dw \Rightarrow dw = \frac{dt}{\sqrt{2}}$. Hence

$$\begin{aligned}
f(z) &= \frac{e^{-z^2/4}}{2\pi\sqrt{2}}\int_{-\infty}^{\infty} e^{-\frac{1}{2}t^2}dt \\
&= \frac{e^{-\frac{1}{2}(\frac{z}{\sqrt{2}})^2}}{2\pi\sqrt{2}}
\end{aligned}$$

$\Rightarrow Z \sim N(0,2)$.

Example 2
Let X_1 and X_2 be a random sample of size 2 from the distribution having

Chapter 8

probability density function

$$f(x) = \begin{cases} e^{-x} & 0 < x < \infty \\ 0 & \text{elsewhere} \end{cases}$$

given that $Y_1 = X_1 + X_2$ and $Y_2 = \dfrac{X_1}{X_1 + X_2}$, show that Y_1 and Y_2 are independent.

Solution
Since (X_1, X_2) is a random sample,

$$f(x_1, x_2) = e^{-(x_1+x_2)} \quad 0 < x_2, x_2 < \infty.$$

Since
$$y_1 = x_1 + x_2 \tag{1}$$

and
$$y_2 = \dfrac{x_1}{x_1 + x_2} \tag{2}$$

$$\therefore y_2 = \dfrac{x_1}{x_1 + x_2} = \dfrac{x_1}{y_1} \quad \therefore x_1 = y_1 y_2 \tag{3}$$

From (1), $x_2 = y_1 - x_1 \Rightarrow x_2 = y_1 - y_1 y_2$.

$$J = \dfrac{\partial(x_1, x_2)}{\partial(y_1, y_2)} = \begin{vmatrix} \dfrac{\partial x_1}{\partial y_1} & \dfrac{\partial x_1}{\partial y_2} \\ \dfrac{\partial x_2}{\partial y_1} & \dfrac{\partial x_2}{\partial y_2} \end{vmatrix}$$

$$= \begin{vmatrix} y_2 & y_1 \\ 1 - y_2 & -y_1 \end{vmatrix}$$

$$= |-y_1 y_2 - y_1(1 - y_2)|$$

$$= |-y_1 y_2 - y_1 + y_1 y_2| = |-y_1|$$

Chapter 8

$$\therefore |J| = |-y_1| = y_1$$
$$g(y_1, y_2) = e^{-(y_1 y_2 + y_1 - y_1 y_2)}|J|$$
$$\therefore g(y_1, y_2) = e^{-(y_1 y_2 + y_1 - y_1 y_2)} y_1$$
$$= y_1 e^{-(y_1 y_2 + y_1 - y_1 y_2)} = y_1 e^{y_1}$$
$$= h(y_1) \ k(y_2)$$

where $h(y_1) = y_1$ and $k(y_2) = e^{-y_1}$.
Hence y_1 and y_2 are independent.

Discrete Random Variables.

Suppose X and Y are independent discrete random variables with probability density functions $f(x)$ and $g(y)$ respectively, if
$W = X + Y$ and $X = R$ then $h(w) = \sum_{r=0}^{W} f(r) \ g(w - r)$.
The above is convolution theorem for independent discrete random variables.

Example 1

Given that X is a discrete random variable which has a Poisson distribution with mean λ and Y is also a discrete random variable which has Poisson distribution with mean μ, find the distribution of $W = X + Y$. [X, Y are independent].

Solution

Let $X = R \Rightarrow x = r$ \hfill (1)
$\therefore W = x + y \Rightarrow w = r + y \Rightarrow y = w - r$ \hfill (2)
Since X has a Poisson distribution

$$\therefore f(x) = \frac{\lambda^x e^{-\lambda}}{x!}$$

Since Y also has Poisson distribution

$$\therefore f(x) = \frac{\lambda^y e^{-\lambda}}{y!}$$

Chapter 8   199

$h(x) = P(W = w)$, from (1) and (2), $x = r$, $y = w - r$.
By convolution, w

$$\therefore \phi(W) = \sum_{r=0}^{W} f(r)g(w-r) \qquad (x \text{ and } y \text{ are independent})$$

$$= \sum_{r=0}^{W} \frac{\lambda^r e^{-\lambda}}{r!} \cdot \frac{\mu^{w-r} e^{-\mu}}{(w-r)!}$$

$$= \frac{e^{-(\lambda+\mu)}}{w!} \sum_{r=0}^{W} \frac{\lambda^r \mu^{w-r} w!}{r!(w-r)!}$$

$$= \frac{e^{-(\lambda+\mu)}}{w!} \sum_{r=0}^{W} \frac{w!}{r!(w-r)!} \lambda^r \mu^{w-r}$$

$$= \frac{e^{-(\lambda+\mu)}}{w!} \sum_{r=0}^{W} \binom{w}{r} \lambda^r \mu^{w-r}$$

$$= \frac{(\lambda+\mu)^w e^{-(\lambda+\mu)}}{w!}$$

$$= \frac{e^{-(\lambda+\mu)}}{w!} (\lambda+\mu)^w$$

This is a Poisson distribution with mean $\lambda + \mu$.

Example 2
Given that X is a discrete random variable from Binomial distribution with parameter n_1, θ and Y is another discrete random variable from Binomial distribution with parameters n_2, θ. If X and Y are independent, find the probability density function of $W = X + Y$.

Solution
Since X is from Binomial distribution with parameters (n_1, θ),

$$f(x) = \binom{n_1}{x} \theta^x (1-\theta)^{n_1-x}.$$

Similarly,
$$f(y) = \binom{n_2}{y} \theta^y (1-\theta)^{n_2-y}.$$

Let $W = x + y$ and $x = r$.

$$\therefore \phi(w) = \sum_{r=0}^{w} f(r)g(w-r)$$

$$= \sum_{r=0}^{w} \binom{n_1}{r} \theta^r (1-\theta)^{n_1-r} \binom{n_2}{w-r} \theta^{w-r} (1-\theta)^{n_2-(w-r)}$$

$$= \sum_{r=0}^{w} \binom{n_1}{r} \binom{n_2}{w-r} \theta^{r+w-r} (1-\theta)^{n_1-r+n_2-w+r}$$

$$\therefore \phi(w) = \sum_{r=0}^{w} \binom{n_1}{r} \binom{n_2}{w-r} \theta^w (1-\theta)^{n_1+n_2-w}$$

This is binomial distribution with parameters $(n_1 + n_2, \theta)$.

Moment Generating Function/Characteristic Function Approach
Example 1
Given that the distribution of χ_i is $\chi^2_{(n_i)}$, $i = 1, 2, 3, \ldots, k$ where X_i's are independent random variables.
If $Z = X_1 + X_2 + X_3 + \cdots + X_k$, show that $Z \sim \chi^2(n)$ where $n = n_1 + n_2 + \cdots + n_k$.
(**Hint** Use moment generating function approach).

Proof:
For a chi-square distribution,
$$M_{X_i}(t) = (1-2t)^{-n_i/2}$$

where $M_X(t)$ is the moment generating function of

$$\begin{aligned} M_Z(t) &= M_{\sum X_i}(t)(1-2t) \\ &= (1-2t)^{-n_1/2} \cdot (1-2t)^{-n_2/2} \cdots (1-2t)^{-n_k/2} \\ &= (1-2t)^{-\frac{1}{2}(n_1+n_2+\cdots+n_k)} \\ &= (1-2t)^{-\frac{n}{2}} \end{aligned}$$

Chapter 8

$\Rightarrow Z \sim \chi^2_{(n)}$ proved.

Example 2
Given that X is a random variable from Poisson distribution with mean λ and Y is Poisson distribution with mean μ, use characteristic function approach to find the probability density function of $W = X + Y$.

Solution:
Characteristic function of $X = \phi_X(t) = e^{\lambda(e^{it}-1)}$
Characteristic function of $Y = \phi_Y(t) e^{\mu(e^{it}-1)}$.
Since X and Y are independent

$$\begin{aligned}\phi_W(t) &= \phi_{X+Y}(t) = \phi_X(t) \times \phi_Y(t) = e^{\lambda(e^{it}-1)} \times e^{\mu(e^{it}-1)} \\ &= e^{(\lambda+\mu)(e^{it}-1)}\end{aligned}$$

This is a characteristic function of a Poisson distribution with mean $(\lambda + \mu)$. Hence by uniqueness theorem

$$\phi(W) = \frac{(\lambda+\mu)^w e^{-(\lambda+\mu)}}{w!},$$

where $w = x + y$.

Example 3
Given that $X_1, X_2, X_3, \ldots, X_n$ are continuous independent random variables from exponential distribution each with probability density function
$f(x) = \lambda e^{-\lambda x} \quad x > 0$
Using characteristic function approach, find the probability density function of $W = \sum_{k=1}^{n} X_k$.

Solution

$$\phi_X(t) = \left(1 - \frac{it}{\lambda}\right)^{-1}$$

Chapter 8

Since

$$W = \sum_{k=1}^{n} X_k, \quad \phi_W(t) = \phi_{\sum X_k}(t)$$

$$= \left(1 - \frac{it}{\lambda}\right)^{-1} \left(1 - \frac{it}{\lambda}\right)^{-1} \left(1 - \frac{it}{\lambda}\right)^{-1} \cdots \left(1 - \frac{it}{\lambda}\right)^{-1}$$

$$= \left(1 - \frac{it}{\lambda}\right)^{-n}$$

$\phi_W(t)$ is the characteristic function of a $\Gamma(n)$ distribution with parameter λ,
$$h(w) = \frac{\lambda(\lambda w)^{n-1} e^{-\lambda w}}{\Gamma(n)}, \quad \lambda > 0, \quad w > 0.$$

Example 4
Let X_1 and X_2 be stochastically independent with normal distributions $N(\mu_1, \sigma^2)$ and $N(\mu_2, \sigma^2)$ respectively. Define the random variable $Y = X_1 - X_2$. Use moment generating function approach to find the distribution of Y.

Solution:
By definition

$$\begin{aligned}
M_y(t) &= E(e^{ty}) \\
&= E(e^{t(x_1 - x_2)}) \\
&= E(e^{tx_1 - tx_2}) \\
&= E(e^{tx_1} e^{-tx_2}) \\
&= E(e^{tx_1}) \cdot E(e^{tx_2}) \quad (\because X \text{ and } Y \text{ are independent}) \\
&= M_{X_1}(t) \cdot M_{X_2}(-t)
\end{aligned}$$

But for $X \sim N(\mu, \sigma^2)$
$$M_X(t) = e^{\mu t + \frac{1}{2}\sigma^2 t^2}$$

$\therefore$ For $X_1 \sim N(\mu_1, \sigma_1^2)$,
$$M_{X_1}(t) = e^{\mu_1 t + \frac{1}{2}\sigma_1 t^2}$$

For $X_2 \sim N(\mu_2, \sigma_2^2)$
$$M_{X_2}(t) = e^{-\mu_2 t + \frac{1}{2}\sigma_2 t^2}$$

Chapter 8

$$\begin{aligned}
\therefore \quad M_Y(t) = E(e^{tY}) &= M_{X_1}(t) \cdot M_{X_2}(-t) \\
&= e^{\mu_1 t + \frac{1}{2}\sigma_1^2 t^2} \cdot e^{\mu_2(-t) + \frac{1}{2}\sigma_2^2(-t)^2} \\
&= e^{\mu_1 t + \frac{1}{2}\sigma_1^2 t^2} \cdot e^{-\mu_2 t + \frac{1}{2}\sigma_2^2 t^2} \\
&= e^{\mu_1 t + \frac{1}{2}\sigma_1^2 t^2 - \mu_2 t + \frac{1}{2}\sigma_2^2 t^2} \\
&= e^{(\mu_1 - \mu_2)t + \frac{1}{2}(\sigma_1^2 + \sigma_2^2)t^2}
\end{aligned} \qquad (1)$$

Recall that for $X \sim N(\mu, \sigma^2)$

$$M_X(t) = e^{\mu t + \frac{1}{2}\sigma^2 t^2} \qquad (2)$$

Comparing (1) with (2), $\mu = \mu_1 - \mu_2$, $\sigma^2 = \sigma_1^2 + \sigma_2^2$

$$\therefore \quad Y \sim N(\mu_1 - \mu_2, \ \sigma_1^2 + \sigma_2^2),$$

Miscellaneous Examples
Example 1:
Given that $X_1, X_2, \ldots, X_n$ denote a random sample size $n \geq 2$ from a distribution which is $N(\mu, \sigma^2)$. Find the distribution of $\bar{X} = \frac{1}{n}\sum x_i$.

Solution
Let $Y = \bar{X}$

$$\begin{aligned}
E(Y) = E(\bar{X}) &= E\left(\frac{1}{n}\sum x_i\right) \\
&= \frac{1}{n}E(\sum x_i) \\
&= \frac{1}{n}\sum E(X_i) \\
&= \frac{1}{n}\sum \mu \\
&= \frac{1}{n} \cdot n\mu = \mu
\end{aligned}$$

Chapter 8

$$Var(Y) = Var(\bar{X}) = Var\left(\frac{1}{n}\sum x_i\right)$$
$$= \frac{1}{n^2}\sum Var(X_i)$$
$$= \frac{1}{n^2}\sum \sigma^2$$
$$= \frac{n\sigma^2}{n^2} = \frac{\sigma^2}{n}$$

∴ The distribution is $\bar{X} \sim \left(\mu, \frac{\sigma^2}{n}\right)$.

Example 2:
Given that X is normally distributed with mean 30 and variance 16, (a) obtain (i) $P(X \leq 20)$ (ii) $P(15 \leq X \leq 25)$. (b) Using the same information for a random sample of size $n = 25$, obtain $P(\bar{X} \leq 28)$.

Solution
Since $X \sim N(0, 1)$,
$$f(x) = \frac{1}{\sqrt{2\pi}}e^{-x^2/2}$$

$$M_y(t) = E(e^{ty}) = \frac{1}{\sqrt{2\pi}}\int_{-\infty}^{\infty}e^{ty}e^{-x^2/2}dx$$
$$= \frac{1}{\sqrt{2\pi}}\int_{-\infty}^{\infty}e^{tx^2}e^{-x^2/2}dx$$
$$= \frac{1}{\sqrt{2\pi}}\int_{-\infty}^{\infty}e^{-x^2/2(1-2t)}dx$$

Set $w = \frac{x^2}{2}(1-2t)$ \hfill (2)

$$\frac{dw}{dx} = \frac{2x}{2}(1-2t) \Rightarrow dw = x(1-2t)dx$$

$$\therefore dx = \frac{dw}{x(1-2)} \hfill (3)$$

Chapter 8

From (2),
$$x^2 = \frac{2w}{1-2t} \Rightarrow x = \left(\frac{2w}{1-2t}\right)^{\frac{1}{2}} \qquad (4)$$

Substitute (2), (3) and (4) into (1)

$$\begin{aligned}
\therefore \quad M_y(t) = E(e^{ty}) &= \frac{1}{\sqrt{2\pi}} \int_{-\infty}^{\infty} e^{-w} \frac{1}{x(1-2t)} dw \\
&= \frac{1}{\sqrt{2\pi}} \int_{-\infty}^{\infty} e^{-w} \frac{x^{-1}}{1-2t} dw \\
&= \frac{1}{\sqrt{2\pi}} \int_{-\infty}^{\infty} e^{-w} \left(\frac{2w}{1-2t}\right)^{-\frac{1}{2}} \left(\frac{1}{1-2t}\right) dw \\
&= \frac{1}{\sqrt{2\pi}} \int_{-\infty}^{\infty} e^{-w} w^{-\frac{1}{2}} \frac{2^{-\frac{1}{2}}}{(1-2t)^{-\frac{1}{2}}(1-2t)} dw \\
&= \frac{1}{\sqrt{2\pi}} \int_{-\infty}^{\infty} e^{-w} w^{-\frac{1}{2}} \frac{1}{\sqrt{2}(1-2t)^{\frac{1}{2}}} dw \\
&= \frac{1}{2\sqrt{\pi(1-2t)}} \int_{-\infty}^{\infty} e^{-w} w^{-\frac{1}{2}} dw
\end{aligned}$$

$$\begin{aligned}
\therefore \quad M_y(t) &= \frac{(1-2t)^{-\frac{1}{2}}}{2\sqrt{\pi}} \int_{-\infty}^{\infty} e^{-w} w^{\frac{1}{2}-1} dw \\
&= (1-2t)^{-\frac{1}{2}} \int_{-\infty}^{\infty} \frac{1}{2\sqrt{\pi}} e^{-w} w^{\frac{1}{2}-1} dw \\
&= (1-2t)^{-\frac{1}{2}} \frac{\Gamma(\frac{1}{2})}{\sqrt{\pi}} \\
&= (1-2t)^{-\frac{1}{2}} \frac{\sqrt{\pi}}{\sqrt{\pi}} \quad \text{(since } \Gamma(\frac{1}{2}) = \sqrt{\pi} \\
&= (1-2t)^{-\frac{1}{2}} \text{ which is } \chi^2(1)\text{'}
\end{aligned}$$

Chapter 9

Central Limit Theorem and Application

Statement and Proof
Statement

The theorem states that, if $f(\cdot)$ is a density function with finite mean μ and finite variance σ^2 and $\bar{X}_n$ is the sample mean where $X_1, X_2, \ldots, X_n$, are independent and identically distributed (i.i.d) then the random variable

$$Z_n = \frac{\bar{X}_n - E(\bar{X}_n)}{\sqrt{Var(\bar{X}_n)}}$$

is a standardized variable and is asymptotically normal, i.e.

$$\lim_{n \to \infty} M_{Z_n}(\theta) = M_Z(\theta) = e^{\frac{1}{2}\theta^2}$$

where

$$\bar{X}_n = \frac{\sum_{i}^{n} x_i}{n} \quad i = 1, 2, \ldots, n$$

Chapter 9

Proof

(i) $$\begin{aligned}Z_n &= \frac{\bar{X}_n - E(\bar{X}_n)}{\sqrt{Var(\bar{X}_n)}} \\ &= \frac{\bar{X}_n - \mu}{\sigma/\sqrt{n}} \\ &= \frac{(\bar{X}_n - \mu)\sqrt{n}}{\sigma} \\ &= \frac{\left(\sum x_i - n\mu\right)\sqrt{n}}{n\sigma} \quad i = 1, 2, \ldots, n \\ &= \frac{\sum x_i - n\mu}{\sigma\sqrt{n}}\end{aligned}$$

It then follows from above that Z_n is a standardized variable associated with $\bar{X}$.

(ii) Recall that given $Z \sim N(0,1)$, the moment generating function of Z is denoted by $M_Z(\theta) = E(e^{\theta Z}) = e^{\frac{1}{2}\theta^2}$.

Let $M_{Z_n}(\theta)$ denote the moment generating function of Z_n, we want to show that
$$\lim_{n \to \infty} M_{Z_n}(\theta) = M_Z(\theta) = e^{\frac{1}{2}\theta^2}$$

Proof

$$\begin{aligned}M_{Z_n}(\theta) &= E(e^{\theta Z_n}) = E[\exp(\theta Z_n)] \\ &= E\left[\left(\exp\theta\left(\frac{\sum x_i - n\mu}{\sigma\sqrt{n}}\right)\right)\right] \\ &= E\left\{\exp\left[\frac{\theta}{n}\sum_{i=1}^{n}\left(\frac{x_i - \mu}{\sigma/\sqrt{n}}\right)\right]\right\} \\ &= E\left[\prod_{i=1}^{n}\exp\left(\frac{\theta}{n}\frac{x_i - \mu}{\sigma/\sqrt{n}}\right)\right]\end{aligned}$$

Since the random variables are mutually independent, then

$$M_{Z_n}(\theta) = \prod_{i=1}^{n} E\left\{\exp\left(\frac{\theta}{n}\frac{x_i-\mu}{\sigma/\sqrt{n}}\right)\right\}$$

$$= \prod_{i=1}^{n} E\left\{\exp\left[\frac{\theta}{\sqrt{n}}\left(\frac{x_i-\mu}{\sigma}\right)\right]\right\}$$

Since the random variables are also identically distributed

$$M_{Z_n}(\theta) = \left\{E\left[\exp\left(\frac{\theta}{\sqrt{n}}\frac{x_i-\mu}{\sigma}\right)\right]\right\}^n ; \quad \text{set } Y = x_i - \mu; \quad i = 1, 2, 3, \ldots, n.$$

Since all the random variables are identically distributed.

$$Y = \frac{x-\mu}{\sigma} \; \forall \; i$$

$$M_{Z_n}(\theta) = E\left[(\exp\frac{\theta}{\sqrt{n}}Y)\right]^2$$

$$= \left[M_Y \frac{\theta}{\sqrt{n}}\right]^n$$

$$= \left[M_Y\left(\frac{\theta}{\sqrt{n}}\right)\right]^n$$

where $M_Y\left(\frac{\theta}{\sqrt{n}}\right)$ is the common m.g.f. of each

$$Y_i = \frac{x_i-\mu}{\sigma}, \quad i = 1, 2, \ldots, n.$$

Since $M_Y\left(\frac{\theta}{\sqrt{n}}\right) = E\left[\exp\left(\frac{\theta}{\sqrt{n}}Y\right)\right]$ is the common m.g.f.

$$M_Y\left(\frac{\theta}{\sqrt{n}}\right) = E\left[\exp\left(\frac{\theta}{\sqrt{n}}Y\right)\right] = E\left[1 + \frac{\theta}{\sqrt{n}}Y + \frac{1}{2!}\frac{\theta^2}{n}Y^2 + \frac{1}{3!}\frac{\theta^3}{(\sqrt{n})^3}Y^3 + \cdots\right]$$

$$= 1 + \frac{\theta}{\sqrt{n}}E(Y) + \frac{\theta^2}{2n}E(Y^2) + \frac{\theta^3}{6n\sqrt{n}}E(Y^3) + \cdots$$

$$= 1 + \frac{\theta^2}{\sqrt{n}}(0) + \frac{\theta^2}{2n}\left(\frac{\sigma^2}{\sigma^2}\right) + \frac{\theta^3\mu_3}{3!(\sqrt{n})^3} + \cdots \quad \because \left(Y = \frac{x-\mu}{\sigma}\right)$$

Chapter 9

$$= 1 + \frac{\theta^2}{2n} + \frac{\theta^3}{3!(\sqrt{n})^3}\mu_3 + \cdots$$

$$= 1 + \frac{1}{n}\left(\frac{1}{2}\theta^2 + \frac{\theta^3}{3!\sqrt{n}}\mu_3 + \cdots\right)$$

$$= 1 + \frac{v}{n} \quad \left(\text{where} \quad v = \frac{1}{2}\theta^2 + \frac{\theta^3}{3!\sqrt{n}}\mu_3 + \cdots\right)$$

Hence,

$$M_{Z_n}(\theta) = \left[M_Y\left(\frac{\theta}{\sqrt{n}}\right)\right]^n = \left(1 + \frac{v}{n}\right)^n$$

$$\lim_{n\to\infty} M_{Z_n}(\theta) = \lim_{n\to\infty}\left(1 + \frac{v}{n}\right)^n = e^{v^*}$$

where $v^* = \lim_{n\to\infty} v = \left(\frac{1}{2}\theta^2 + \frac{\theta^2}{3!}\frac{\mu_3}{\sqrt{n}} + \cdots\right) = \frac{1}{2}\theta^2$.

So that $\lim_{n\to\infty} M_{Z_n}(\theta) = e^{\frac{1}{2}\theta^2}$ which is the m.g.f. of a standardized normal with zero mean and unit variance.

Hence, $\lim_{n\to\infty} M_{Z_n}(\theta) = M_Z(\theta) = e^{\frac{1}{2}\theta^2}$ \qquad (Proved).

Application of Central Limit Theorem

Let $X_1, X_2, \ldots,$ be independent and have a common distribution with finite mean μ and finite variance σ^2. Then the distribution of their sum $S_n = X_1 + X_2 + \cdots + X_n$ is distributed approximately as $N(n\mu, n\sigma^2)$, and probabilities can be approximated as $P(S_n \leq y) = \phi\left(\frac{y - n\mu}{\sqrt{n\sigma^2}}\right)$, where y is a particular value of S_n. The larger the sample size n, the better the approximation, but no general rule can be given as to how large n should be. A sufficiently large value of n may require $n > 25$.

However, if the underlined distribution is symmetrical and unimodal and of continuous type a value of n as small as 4 may be enough for an adequate approximation particularly if the underlined distribution is normal that is $S_n \sim N\left(\mu, \frac{\sigma^2}{n}\right)$ for $n = 1, 2, \ldots$.

Example 1
An elevator sign reads "maximum weight 2700 tonnes, capacity 17 passengers".

Chapter 9   210

If the population of passengers has mean 150 tonnes and standard deviation 25 tonnes, what is the probability that a capacity crowd will overload the elevator?

Solution

The total weight of 17 passengers is a random variables with mean 150 tonnes $= 17 \times 150 = 2550$ and standard deviation $= \sigma\sqrt{n} = 25\sqrt{17} = 103$. Let X be a r.v. denoting the total weight

$$
\begin{aligned}
P(X > 2700) &= 1 - P(X < 2700) \\
&= 1 - \phi\left(\frac{2700 - 2550}{103}\right) \\
&= 1 - \phi(1.46) \\
&= 1 - 0.9279 \\
&= 0.0721
\end{aligned}
$$

Example 2

Given that X is normally distributed with mean 30 and variance 16, (a) obtain (i) $P(X \le 20)$ (ii) $P(15 \le X \le 25)$. (b) Using the same information for a random sample of size $n = 25$, obtain $P(\bar{X} \le 28)$.

$X \sim N(30, 16)$

$$
\begin{aligned}
\text{(a)(i)} \quad P(X \le 20) &= P\left(\frac{X - \mu}{\sigma} \le \frac{20 - 30}{4}\right) \\
&= P(Z \le -2.5) \\
&= 1 - P(Z \le 2.5) \\
&= 1 - 0.9938 = 0.0062
\end{aligned}
$$

$$
\begin{aligned}
\text{(ii)} \quad P(15 \le X \le 25) &= P\left(\frac{15 - 30}{4} \le \frac{X - \mu}{\sigma} \le \frac{25 - 30}{4}\right) \\
&= P(-3.75 \le Z \le -1.25) \\
&= \phi(-1.25) - \phi(-3.75) \\
&= \phi(3.75) - \phi(1.25) \\
&= 0.9999 - 0.8944 \\
&= 0.1055
\end{aligned}
$$

Chapter 9

(b) (i) Since $n = 25$, $\sigma = 4$ then $\sigma_{\bar{X}} = \dfrac{\sigma}{\sqrt{n}} = \dfrac{4}{\sqrt{25}} = \dfrac{4}{5} = 0.8$

$$\begin{aligned} P(\bar{X} \leq 28) &= P\left(\dfrac{\bar{X} - \mu}{\sigma_{\bar{X}}} \leq \dfrac{28 - 30}{0.8}\right) \\ &= P(Z \leq -2.5) \\ &= 1 - \phi(2.5) \\ &= 1 - 0.9938 \\ &= 0.0062 \end{aligned}$$

Example 3
The mean number of accident per day in a given city is 10 with a standard deviation of 2. Compute the probability that over a period of 16 days an observer will record a mean accident rate of (i) 11 per day or less (ii) between 9 and 11 per day.

Solution
$n = 16$, $\sigma^2 = 4$, $\bar{X} \leq 11$, $\mu = 10$.

$$\begin{aligned} \text{(i)} \quad P(\bar{X} \leq 11) = P\left(\dfrac{\bar{X} - \mu}{\sigma/\sqrt{n}}\right) \leq \dfrac{11 - 10}{2/\sqrt{16}} &= P\left(Z \leq \dfrac{11 - 10}{2/4}\right) \\ &= P(Z \leq 2) = 0.9772 \end{aligned}$$

Here $\bar{X}_1 = 9$, $\bar{X}_2 = 11$

$$\begin{aligned} \text{(ii)} \quad Pr(9 \leq \bar{X} \leq 11) &= P\left(\dfrac{\bar{X}_1 - \mu}{\sigma/\sqrt{n}}\right) \leq Z \leq \dfrac{P(\bar{X}_2 - \mu)}{\sigma/\sqrt{n}} \\ &= P\left(\dfrac{4 - 10}{2/4} \leq Z \leq \dfrac{11 - 10}{2/4}\right) \\ &= P(-2 \leq Z \leq 2) = 2P(Z \leq 2) - 1 \\ &= 2(0.9772) - 1 \\ &= 1.9554 - 1 = 0.9554 \end{aligned}$$

Chapter 9

Importance of Central Limit Theorem

(1) It is an important theorem that gives asymptotic or limiting distribution of $\bar{X}_n$ where $\bar{X}_n$ is the sample mean of a random sample of size n.

(2) It can be used for approximating certain probability concerning the mean $\bar{X}_n$ of a random sample when n is sufficiently large
$$\bar{X}_n \sim N\left(\mu, \frac{\sigma^2}{n}\right) \quad n = 1, 2, 3, \ldots$$

(3) It gives the normal distribution its place in the theory of sampling.

(4) In estimating the error which may arise in a sample, we assume that the distribution of the sample means is normal, that is, the Central Limit Theorem is true.

Markov's Inequality

If X is a continuous random variable that takes only non-negative values then for any value $a > 0$
$$P(X \geq a) \leq \frac{E(X)}{a}$$

Proof:

If X is continuous with $f(x)$ as the density function,

$$\begin{aligned}
E(X) &= \int_0^\infty x f(x) dx \\
&= \int_0^a x f(x) dx + \int_a^\infty x f(x) dx \\
&\geq \int_a^\infty x f(x) dx \\
&\geq \int_a^\infty a f(x) dx \\
&= a \int_a^\infty f(x) dx \\
&= a P(X \geq a)
\end{aligned}$$

$\therefore\ E(X) \geq aP(X \geq a)$

$\therefore\ P(X \geq a) \leq \dfrac{E(X)}{a}$ proved.

Chebyshev's Inequality
Statement

If μ, σ^2 are mean and variance of a r.v. X respectively, then any positive constant k,

$$P(|x - \mu| > k\sigma) \leq \frac{1}{k^2} \quad \text{or} \quad P(|x - \mu| > \epsilon) \leq \frac{\sigma^2}{\epsilon}$$

where ϵ is positive.

Proof from the first principle

Assume that X is continuous.

$$\sigma^2 = \int_{-\infty}^{\infty} (x - \mu)^2 f(x) dx, \text{ by definition}$$

$$= \int_{-\infty}^{\mu - k\sigma} (x - \mu)^2 f(x) dx + \int_{\mu - k\sigma}^{\infty} (x - \mu) f(x) dx$$

If $|x - \mu| > k\sigma$, then $(x - \mu)^2 > k^2 \sigma^2$.
From above,

$$\sigma^2 > k^2 \sigma^2 \int_{-\infty}^{\mu - k\sigma} f(x) dx + k^2 \sigma^2 \int_{\mu + k\sigma}^{\infty} f(x)(dx$$

$$\frac{\sigma^2}{k^2 \sigma^2} > \int_{-\infty}^{\mu - k\sigma} f(x) dx + \int_{\mu + k\sigma}^{\infty} f(x) dx = P(X \leq \mu - k\sigma) + P(X > \mu + k\sigma)$$

Consider the interval $|x - \mu| \geq k\sigma$

$$\therefore \quad -k\sigma \geq x - \mu \geq k\sigma$$

i.e. $\mu - k\sigma \geq x \geq \mu + k\sigma.$

Thus, $x \leq \mu - k\sigma$ or $x \geq \mu - k\sigma$.
Hence, $\frac{1}{k^2} \geq P(x \leq \mu - k\sigma) + P(x \geq \mu + k\sigma) = P(|x - \mu| \geq k\sigma)$

$\therefore \quad P(|x-\mu| > k\sigma) \leq \frac{1}{k^2}$ or $P(|x-\mu| > \epsilon) \leq \frac{\sigma^2}{\epsilon}$ (substituting $k\sigma$ for ϵ). (q.e.d.)

Chapter 9

Proof of Chebyshev's inequality using Markov's inequality approach

Recall that Chebyshev's inequality states that if X is a random variable with mean μ and variance σ^2, then for any value $k > 0$, $P[|X - \mu| \geq k] \leq \dfrac{\sigma^2}{k^2}$.

Proof: Since $(X - \mu)^2$ is a non-negative random variable, we can apply Markov's inequality (with $a = k^2$ to obtain

$$P\{(X - \mu)^2 \geq k^2\} \leq \frac{E(X - \mu)^2}{k^2} \tag{1}$$

Since $E(X - \mu)^2 = \sigma^2$ and $(X - \mu)^2 \geq k^2$ iff $|X - \mu| > k$ then (1) is equivalent to

$$P\{|X - \mu| \geq k\} \leq \frac{E(X - \mu)^2}{k^2} = \frac{\sigma^2}{k^2}$$

$$\therefore P\{|X - \mu| \geq k\} \leq \frac{\sigma^2}{k^2}$$

or

$$P\{|X - \mu| < k\} > 1 - \frac{\sigma^2}{k^2}.$$

Importance of the two inequalities:

Markov's inequality enables us to derive bounds on probabilities when only the mean of the probability distribution is given. Chebyshev's inequality enables us to set bounds on probabilities when both the mean and variance of the probability distribution are known.

Other forms of Chebyshev's Inequality

Put $k\sigma = \lambda$ then $k = \dfrac{\lambda}{\sigma}$ and $\dfrac{1}{k} = \dfrac{\sigma}{\lambda}$

(1) $P(|x - \mu| > \lambda) \leq \dfrac{\sigma^2}{\lambda^2}$

(2) $P(|x - \mu| < k\sigma) \geq 1 - \dfrac{1}{k^2}$

(3) $P(|x - \mu| < \lambda) \geq 1 - \dfrac{\sigma^2}{\lambda^2}$

Application of Chebyshev's Inequality
Example 1

A random variable X has $\mu = 12$, $\sigma^2 = 9$ and an unknown probability distribution, using Chebyshev's inequality, find
(a) $P(6 < x < 18)$ (b) $P(3 < x < 21)$.

Solution

(a) Since $\mu = 12$.
$\therefore \ P(6 < x < 18) = P(6 - 12 < x - 12 < 18 - 12)$
$= P|x - 12| < 6$.
Chebyshev's inequality states that given μ and σ^2 and +ve constant k,

$$P(|x - \mu| > k\sigma) \leq \dfrac{1}{k^2} \qquad (*)$$

$\mu = 12$, $\sigma^2 = 9$, $\sigma = 3$.
Substitute for μ and σ in $(*)$, we obtain

$$P(|x - 12| > 3k) \leq \dfrac{1}{k^2}$$

$$\therefore \ 3k = 6 \Rightarrow k = 2$$

Hence, $P(|x - 12| > 6) \leq \dfrac{1}{4}$.
But $P(|x - 12| < 6) = 1 - P(|x - 12| > 6)$

$$\therefore \ P(|x - 12| < 6) \geq 1 - \dfrac{1}{4} = \dfrac{3}{4}$$

$$\therefore \ P(6 < x < 18) = \dfrac{3}{4}$$

Chapter 9

(b) $\quad P(3 < x < 21) = P(3 - 12 < x - 12 < 21 - 12)$
$= P(-9 < x - 12 < 9) = P(|x - 12|) < 9$
$k\sigma = 9, \ \sigma = 3 \Rightarrow k = 3$

$\therefore \ P(|x - 12| > 9) \leq \dfrac{1}{k^2} = \dfrac{1}{9}$

$P(|x - 12| < 9) = 1 - \dfrac{1}{9} = \dfrac{8}{9}$

$\therefore \ P(3 < x < 21) \geq \dfrac{8}{9}$

Example 2
A random variable X has $\mu = 10$, $\sigma^2 = 4$, using Chebyshev's inequality, find

(a) $P(|x - 10| \geq 3)$

Solution

(a) $P(x - \mu| > k\sigma) \leq \dfrac{1}{k^2}$ (by Chebyshev's inequality)

$\sigma^2 = 4$; i.e. $\sigma = 2$.
$k\sigma = 3 \Rightarrow 2k = 3$ and $k = \frac{3}{2}$
$P(|x - 10| \geq 3) \leq \dfrac{4}{9}$

Example 3
A random variable X has $E(X) = 5$ and $V(X) = 4$, where $E(X)$ and $V(X)$ are expected value and variance (σ^2) respectively. Using Chebyshev's inequality,

(i) For what value of k will $P(|x - 5| > 2k) < \frac{1}{25}$?

(ii) For what value of h will $P(|X - 5| \leq h) > 0.99$?

Solution

(i) By Chebyshev's inequality

$$P(|x - \mu| > k\sigma| < \dfrac{1}{k^2} \tag{1}$$

Compare (i) with

$$P(|x - 5| > 2k) < \dfrac{1}{25} \quad \text{where } E(X = \mu$$

$$\frac{1}{k^2} = \frac{1}{25} \Rightarrow k^2 = 25 \Rightarrow k = \pm 5$$

But k is positive, therefore $k = 5$.

(ii) Using other form of Chebyshev's inequality:

$$P(|X - \mu| \leq \lambda) > 1 - \frac{\sigma^2}{\lambda^2} \qquad (2)$$

Compare (2) with

$$P(|x - 5| \leq h) > 0.99,$$

$$h = \lambda, \quad 1 - \frac{\sigma^2}{\lambda^2} = 0.99 \qquad (3)$$

Substitute h for λ in (3), we obtain

$$1 - \frac{\sigma^2}{h^2} = 0.99$$

$$1 - \frac{4}{h^2} = 0.99 \quad (\sigma^2 = 4)$$

$$\therefore h^2 = 400$$

$$h = \pm 20$$

since $h > 0$, $h = 20$.

Example 4.

If it is known that the number of items produced in a factory during a week is a random variable with mean 500, what is the probability that the week's production will be at least 1000?

Solution:

$E(X) = \mu = 500$, $a = 1000$. Since only the mean (μ) is known, we shall use Markov's inequality.

By Markov's inequality, $P(X \geq a) \leq \frac{E(X)}{a}$

$$\therefore P(X \geq 1000) \leq \frac{500}{1000} = \frac{1}{2}$$

$$\therefore P(X \geq 1000) \leq \frac{1}{2}.$$

Chapter 9

Law of Large Numbers
The Law of large Numbers is a consequence of Chebyshev's inequality.

Theorem
Let $X_1, X_2, \ldots$ be mutually independent random variables (discrete or continuous), each having finite mean μ and variance σ^2, then if
$S_1 = X_1 + X_2 + \cdots + X_n \quad (n = 1, 2, \ldots)$

$$\lim_{n \to \infty} P \left| \frac{S_n}{n} - \mu \right| \geq \epsilon = 0$$

Since $\dfrac{S_n}{n}$ is the arithmetic mean of $X_1, X_2, \ldots, X_n$, this theorem states that the probability of the arithmetic mean $\dfrac{S_n}{n}$ differing from its expected value μ by more than ϵ approaches zero as $n \to \infty$.

Proof
We know that

$$E(X_1) = E(X_2) = \cdots = E(X_n) = \mu$$
$$Var(X_1) = Var(X_2) = \cdots = Var(X_n) = \sigma^2$$

Then,
$$E\left(\frac{S_n}{n}\right) = E\left(\frac{X_1 + X_2 + \cdots + X_n}{n}\right)$$
$$= \frac{1}{n}[E(X_1) + \cdots + E(X_n)]$$
$$= \frac{1}{n}(n\mu)$$
$$= \mu$$

Similarly, $Var(S_n) = Var(X_1 + X_2 + X_3 + \cdots + X_n) = Var(X_1) + Var(X_2) + \cdots$
Since $Var(X_1) = Var(X_2) = \cdots = Var(X_n)$
$Var(S_n) = n\sigma^2$ so that,

$$Var\left(\frac{S_n}{n}\right) = \frac{1}{n^2} Var(S_n) = \frac{\sigma^2}{n}$$

Recall that by Chebyshev's inequality we have $P(|X - \mu| > \epsilon) \leq \dfrac{\sigma^2}{\epsilon}$.

With $X = \dfrac{S_n}{n}$ and variance $\left(\dfrac{S_n}{n}\right) = \dfrac{\sigma^2}{n}$ we obtain

Taking the limit as $n \to \infty$, this becomes

$$\lim_{n \to \infty} P\left(\left|\dfrac{S_n}{n} - \mu\right| \geq \epsilon\right) \leq \lim_{n \to \infty} \dfrac{\sigma^2}{n\epsilon} = 0$$

$$\therefore \lim_{n \to \infty} P\left(\left|\dfrac{S_n}{n} - \mu\right| \geq \epsilon\right) = 0 \quad \text{proved}$$

Chapter 10

Point Estimation

If we use the value of a statistic to estimate a population parameter, this value is a point estimate of the parameter. For instance, a sample mean can be used to estimate the mean of a population. We can also use a sample variance to estimate the variance of a population. These estimates are called point estimates. They are point estimates because they are single numerical values. The statistic that is used in computing the numerical value is the estimator: For instance in the estimation of the population mean, statistic $\bar{X}$ (the sample mean) is used. Hence $\bar{X}$ is an estimator. In estimating variance of the population the statistic S^2 (sample variance) is used. The statistic S^2 is therefore an estimator.

In summary, a point estimate can be defined as a single observed numerical value used as an estimate of the unknown population parameter.

An estimator is the statistic used in computing the numerical value (point estimate).

It should however be noted that an estimator of a population parameter is obtained from a sample drawn from the population.

For example, if X is a random variable with the following distribution $f(x, \theta) = \theta e^{-\theta x}$ $0 < x < \infty$, θ is the population parameter. An estimator of θ obtained from a sample is denoted by $\hat{\theta}$.

Chapter 10

Properties of a Good Estimator
Unbiasedness

A statistic is said to be unbiased estimator of a population parameter if the mean or expectation of the statistic is equal to the parameter. In other words a statistic T is said to be an unbiased estimator of a parameter θ if $E(T) = \theta$.

Example 1

A random variable X is $B(np, npq)$. Show that X is unbiased for np.

Solution

Required: To show that X is unbiased for np i.e. to show that $E(X) = np$.

Proof: $X \sim B(np, npq) \Rightarrow X$ is a random variable from binomial distribution with mean np and standard deviation $\sqrt{npq}$.

$$f(x) = \begin{cases} \binom{n}{x} p^x q^{n-x} & x = 0, 1, 2, \ldots, n \\ 0 & \text{elsewhere.} \end{cases}$$

$$\therefore E(X) = \sum x f(x)$$

$$= \sum_{x=0}^{n} x \binom{n}{x} p^x q^{n-x}$$

$$= \sum_{x=0}^{n} \frac{x n!}{x!(n-x)!} p^x q^{n-x}$$

$$= \sum_{x=0}^{n} \frac{x n! p^x q^{n-x}}{x(x-1)!(n-x)!}$$

$$= \sum_{x=1}^{n} \frac{n(n-1)!}{(x-1)!(n-x)!} p^x q^{n-x}$$

Chapter 10

$$\begin{aligned}
&= np \sum_{x=1}^{n} \frac{(n-1)! p^{x-1} q^{n-x}}{(x-1)!(n-x)!} \\
&= np \sum_{x=1}^{n} \binom{n-1}{x-1} p^{x-1} q^{n-x} \\
&= np(p+q)^{n-1} \\
&= np.
\end{aligned}$$

Hence $E(X) = np$ and X is unbiased for np proved.

Example 2:
For a random variable $X \sim P(\lambda, \lambda)$, show that X is an unbiased estimator of λ.

Solution

X is an observation from Poisson Distribution with mean λ and standard deviation $\sqrt{\lambda}$.

$\therefore f(x) = \dfrac{\lambda^x e^{-\lambda}}{x!} \quad x = 0, 1, 2, \ldots, \lambda > 0$

$$\begin{aligned}
\therefore E(X) &= \sum x f(x) \\
&= \sum_{x=0}^{\infty} x \left(\frac{\lambda^x e^{-\lambda}}{x!} \right) \\
&= \sum_{x=0}^{\infty} \frac{x \lambda \cdot \lambda^{x-1} e^{-\lambda}}{x(x-1)!} \\
&= \lambda \sum_{x=1}^{\infty} \frac{\lambda^{x-1} e^{-\lambda}}{(x-1)!} \\
&= \lambda \left(1 + \lambda + \frac{\lambda^2}{2!} + \frac{\lambda^3}{3!} + \cdots + \frac{\lambda^r}{r!} + \cdots \right) e^{-\lambda} \\
&= \lambda e^{\lambda} e^{-\lambda} = \lambda e^0 = \lambda.
\end{aligned}$$

$\therefore E(X) = \lambda.$

Hence X is an unbiased estimator of λ.

Chapter 10   224

Example 3
Given that $y_1, y_2, \ldots, y_n$ are independent observations from an exponential distribution with parameter λ, show that $\bar{Y}$ is an unbiased estimator of $\dfrac{1}{\lambda}$.

Solution
Since the given observations are from exponential distribution,
$f(y) = \lambda e^{-\lambda y} \cdot y \geq 0 \ \forall \ y_i \ \ i = 1, 2, \ldots, n, \ E(y_i) = \dfrac{1}{\lambda}$.

$\bar{Y} = \dfrac{1}{n} \sum_{i=1}^{n} y_i.$

$$\therefore \ E(\bar{Y}) = E\left(\dfrac{1}{n} \sum_{i=1}^{n} y_i\right)$$
$$= \dfrac{1}{n} \left(E \sum_{i=1}^{n} y_i\right)$$
$$= \dfrac{1}{n} \sum_{i=1}^{n} E(y_i)$$
$$= \dfrac{1}{n} \sum_{i=1}^{n} \left(\dfrac{1}{\lambda}\right)$$
$$= \dfrac{1}{n} \left(\dfrac{n}{\lambda}\right) = \dfrac{1}{\lambda}$$

$\therefore \ E(\bar{Y}) = \dfrac{1}{\lambda}.$

Hence $\bar{Y}$ is an unbiased estimator of $\dfrac{1}{\lambda}$.

Example 4
For a random variable $X \sim N(0,1)$, show that X is an unbiased estimator of the mean.

Solution
$X \sim N(0,1)$, X is an observation from Standard Normal distribution with

mean 0 and standard deviation 1,

$$\therefore \quad f(x) = \frac{1}{\sqrt{2\pi}} e^{-\frac{1}{2}x^2} \quad -\infty < x < \infty$$

We want to show that $E(X) = 0$.

Proof:

$$\begin{aligned}
E(X) &= \int xf(x)dx \\
&= \int_{-\infty}^{\infty} x \left(\frac{1}{\sqrt{2\pi}} e^{-\frac{1}{2}x^2} \right) dx \\
&= \int_{-\infty}^{\infty} \frac{1}{\sqrt{2\pi}} xe^{-\frac{1}{2}x^2} dx
\end{aligned}$$

Set $u = e^{-\frac{1}{2}x^2} \Rightarrow du = -xe^{-\frac{1}{2}x^2} dx \Rightarrow dx = \dfrac{du}{-xe^{-\frac{1}{2}x^2}}$.

$$\begin{aligned}
\therefore \quad E(X) &= \frac{1}{\sqrt{2\pi}} \int_{-\infty}^{\infty} xe^{-\frac{1}{2}x^2} \cdot \frac{du}{-xe^{-\frac{1}{2}x^2}} \\
&= \frac{1}{\sqrt{2\pi}} \int_{-\infty}^{\infty} -du = \frac{1}{\sqrt{2\pi}} (-u)_{-\infty}^{\infty} \\
&= \frac{1}{\sqrt{2\pi}} \left(-e^{-\frac{1}{2}x^2} \right) \Big|_{-\infty}^{\infty} \\
&= -\frac{1}{\sqrt{2\pi}} \left(e^{-\frac{1}{2}x^2} \right) \Big|_{-\infty}^{\infty} = 0
\end{aligned}$$

$\therefore \quad E(X) = 0 \quad$ proved.

Efficiency

The most efficient estimator among a group of unbiased estimators is the one with the smallest variance. Thus if T_1 and T_2 are two unbiased estimators of θ i.e. if $E(T_1) = E(T_2) = \theta$ and $Var(T_1) < Var(T_2)$, then T_1 is more efficient than T_2.

The relative efficiency of T_1 with respect to T_2 is given by $e = \dfrac{E(T_1 - \theta)^2}{E(T_2 - \theta)^2}$. If T_1 and T_2 are unbiased, then $e = \dfrac{Var(T_1)}{Var(T_2)}$.

Example.
A simple random sample of size n is drawn from a normal population with mean μ and variance σ^2. If the sample mean $(\bar{X})$ and the sample median $(\tilde{X})$ are two estimators of the population mean μ. Obtain the relative efficiency given that the $Var(\tilde{X}) = 1.57\, Var(\bar{X})$.

Solution

$$Var(\bar{X}) = \sigma^2_{\bar{X}} = \frac{\sigma^2}{n}$$

$$Var(\tilde{X}) = \sigma^2_{\tilde{X}} = 1.57\frac{\sigma^2}{n}$$

Relative efficiency $(e) = \dfrac{1.57\sigma^2/n}{\sigma^2/n} = 1.57$.

Sufficient Estimator (Sufficiency)

Definition (1): The statistic $\hat{\theta}$ is a sufficient estimator of the parameter θ if and only if for each value of $\hat{\theta}$, the conditional distribution of the random sample $x_1, x_2, \ldots, x_n$ given $\hat{\theta} = \theta$ is independent of θ.

Definition (2): The statistic $\hat{\theta}$ is a sufficient estimator of the parameter θ if and only if the joint density or probability distribution of the random sample can be factored so that

$$f(x_1, x_2, \ldots, x_n; \theta) = g(\hat{\theta}, \theta)\, h(x_1, x_2, \ldots, x_n)$$

where $g(\hat{\theta}, \theta)$ depends only on $\hat{\theta}$ and θ and $h(x_1, x_2, \ldots, x_n)$ is independent of θ.

Example 1
Let $X_1, X_2, \ldots, X_n$ be independent Bernoulli random variables with parameter

Chapter 10

θ. Show that $Y = X_1 + X_2 + \cdots + X_n$ is sufficient for θ.

Solution

Since the observations are from Bernoulli distribution
$f(x_i, \theta) = \theta^{x_i}(1-\theta)^{1-x_i}$ for $x_i = 0, 1$.
Joint probability density function of

$$\begin{aligned}
x_1, x_2, \ldots, x_n &= f(x_1, x_2, \ldots, x_n, \theta) \\
&= f(x_1, \theta) \cdot f(x_2, \theta) \cdots f(x_n, \theta) \\
&= [\theta^{x_1}(1-\theta)^{1-x_1}][\theta^{x_2}(1-\theta)^{1-x_2}] \cdots [\theta^{x_n}(1-\theta)^{1-x_n}] \\
&= \theta^{\sum x}(1-\theta)^{n-\sum x} \\
&= \theta^y(1-\theta)^{n-y} \qquad \left(y = \sum x\right) \\
&= \dfrac{\theta^y(1-\theta)^{n-y}\binom{n}{y}}{\binom{n}{y}} \\
&= \dfrac{\binom{n}{y}\theta^y(1-\theta)^{n-y}}{\frac{n!}{y!(n-y)!}} \\
&= \binom{n}{y}\theta^y(1-\theta)^{n-y}\left[\dfrac{y!(n-y)!}{n!}\right] = \theta^y(1-\theta)^{n-y}\binom{n}{y}\left(\dfrac{y!(n-y)!}{n!}\right) \\
&= g(y, \theta)\left[\dfrac{y!(n-y)!}{n!}\right]
\end{aligned}$$

which is the probability density function of $g(y, \theta)$ times a constant independent of θ.
Hence y is sufficient for θ.

Example 2

Show that the statistic $\bar{X}$ is a sufficient estimator of the mean μ of normal population with the known variance σ^2.

Solution: For $X \sim (\mu, \sigma^2)$,

$$f(x, \mu, \sigma^2) = \frac{1}{\sigma\sqrt{2\pi}} e^{-\frac{1}{2}\left(\frac{x_i-\mu}{\sigma}\right)^2}$$

$$\therefore f(x_1, x_2, \ldots, x_n, \mu, \sigma^2) = \left(\frac{1}{\sigma\sqrt{2\pi}}\right)^n e^{-\frac{1}{2}\sum_{i=1}^{n}\left(\frac{x_i-\mu}{\sigma}\right)^2}$$

Since

$$\sum_{i=1}^{n}(x_i - \mu)^2 = \sum_{i=1}^{n}[(x_i - \bar{x}) - (\mu - \bar{x})]^2$$

$$= \sum_{i=1}^{n}(x_i - \bar{x})^2 + n(\bar{x} - \mu)^2$$

$$\therefore f(x, \mu, \sigma^2) = \left\{\frac{\sqrt{n}}{\sigma\sqrt{2\pi}e^{-\frac{1}{2}\left(\frac{\bar{x}-\mu}{\sigma/\sqrt{n}}\right)^2}}\right\} \times \left\{\frac{1}{\sqrt{n}}\left(\frac{1}{\sigma\sqrt{2\pi}}\right)^{n-1} e^{-\frac{1}{2}\sum_{i=0}^{n}\left(\frac{x-\bar{x}}{\sigma}\right)^2}\right\}$$

which is the probability density function of μ times a constant independent of μ.

Hence $\bar{X}$ is sufficient for μ.

Consistent Estimator (Consistency)

Definition: The statistic $\hat{\theta}$ is said to be a consistent estimator of the parameter θ if and only if for each positive consistent k, $\lim_{n\to\infty} P(|\hat{\theta} - \theta| \geq k) = 0$ or equivalently, if and only if

$$\lim_{n\to\infty} P(|\hat{\theta} - \theta| < k) = 1$$

The above definition implies that an estimator is consistent if as the sample size increase, the estimator $\hat{\theta}$ converges to the above probabilistic sense to θ.

In addition, it should be noted that the statistic $\hat{\theta}$ is said to be a consistent estimator of the parameter θ if

Chapter 10

(i) θ is unbiased

(ii) $Var(\hat{\theta}) \to 0$ as $n \to \infty$.

Example: Show that the sample variance S^2 is a consistent estimator of σ^2 for random samples from normal distribution $N(\mu, \sigma^2)$.

Solution

To show that S^2 is a consistent estimator of σ^2, it is sufficient to show that

(i) S^2 is an unbiased estimator of σ^2

(ii) $Var(S^2) \to 0$ as $n \to \infty$.

For random sample, the sample variance

$$S^2 = \frac{1}{n-1} \sum_{i=1}^{n} (x_i - \bar{x})^2$$

(i) $E(S^2) = E\left[\dfrac{1}{n-1} \sum_{i=1}^{n} (x_i - \bar{x})^2\right]$

$= \dfrac{1}{n-1} E \sum_{i=1}^{n} (x_i - \bar{x})^2$

$= \dfrac{1}{n-1} \sum_{i=1}^{n} E(x_i - \bar{x})^2$

$= \dfrac{1}{n-1} E\left[\sum_{i=1}^{n} \{(x_i - \mu) - (\bar{x} - \mu)\}^2\right]$

$= \dfrac{1}{n-1} \left[\sum_{i=1}^{n} E(x_i - \mu)^2 - nE\{(\bar{x} - \mu)^2\}\right]$

Chapter 10

But $E(X_i - \mu)^2 = \sigma^2$ and $E(\bar{X} - \mu)^2 = \dfrac{\sigma^2}{n}$

$$\therefore E(S^2) = \dfrac{1}{n-1}\left[\sum_{i=1}^{n} \sigma^2 - n\left(\dfrac{\sigma^2}{n}\right)\right]$$
$$= \dfrac{1}{n-1}[n\sigma^2 - \sigma^2]$$
$$= \dfrac{\sigma^2(n-1)}{n-1} = \sigma^2$$

Hence S^2 is an unbiased estimator of σ^2.
(ii) We want to show that as $n \to \infty$, $Var(S^2) \to 0$.

Proof:
Recall that for random samples of size n from a normal population with variance σ^2, the sampling distribution of S^2 has mean σ^2 and variance $\dfrac{2\sigma^4}{n-1}$.

$$\therefore Var_{n\to\infty}(S^2) = Var_{n\to\infty}\left(\dfrac{2\sigma^4}{n-1}\right) = 0.$$

From (i) and (ii) above, S is a consistent estimator of σ^2 for random samples from normal distribution.

Minimum Variance Unbiased Estimator
To check whether a given unbiased estimator has the smallest possible variance, that is, whether it is a minimum variance unbiased estimator, we make use of the fact that if $\hat{\theta}$ is an unbiased estimator of θ, then the variance of $\hat{\theta}$ must satisfy the inequality;

$$Var(\hat{\theta}) \geq \dfrac{1}{nE\left[\left(\dfrac{\partial \ln f(x)}{\partial \theta}\right)^2\right]}$$

where $f(x)$ is the value of the population density at x, n is the size of the random sample. The inequality is known as Cramer-Rao inequality.

Chapter 10

Definition:
If $\hat{\theta}$ is an unbiased estimator of θ and $Var(\hat{\theta}) = \dfrac{1}{nE\left[\left(\dfrac{\partial \ln f(x)}{\partial \theta}\right)^2\right]}$, then $\hat{\theta}$ is said to be a minimum variance unbiased estimator of θ.

Example 3
Show that $\bar{X}$ is a minimum variance unbiased estimator of the mean(μ) of a normal population.

Solution
From a normal distribution,

$$f(x) = \frac{1}{\sigma\sqrt{2\pi}} e^{-\frac{1}{2}\left(\frac{x-\mu}{\sigma}\right)^2} \quad -\infty < x < \infty$$

$$\ln f(x) = \ln(\sigma\sqrt{2\pi})^{-1} e^{-\frac{1}{2}\left(\frac{x-\mu}{\sigma}\right)^2}$$

$$= -\ln \sigma\sqrt{2\pi} - \frac{1}{2}\left(\frac{x-\mu}{\sigma}\right)^2$$

$$= -\ln \sigma\sqrt{2\pi} - \frac{1}{2\sigma^2}(x-\mu)^2$$

$$\therefore \frac{\partial \ln f(x)}{\partial \mu} = 0 - \frac{1}{2\sigma^2}(-1)(2)(x-\mu)$$

$$= \frac{1}{\sigma^2}(x-\mu)$$

$$= \frac{1}{\sigma}\left(\frac{x-\mu}{\sigma}\right)$$

$$\therefore \left(\frac{\partial \ln f(x)}{\partial \mu}\right)^2 = \frac{1}{\sigma^2}\left(\frac{x-\mu}{\sigma}\right)^2$$

$$\therefore E\left[\left(\frac{\partial \ln f(x)}{\partial \mu}\right)^2\right] = E\left[\frac{1}{\sigma^2}\left(\frac{x-\mu}{\sigma}\right)^2\right]$$

$$= \frac{1}{\sigma^2} E\left(\frac{x-\mu}{\sigma}\right)^2 = \frac{1}{\sigma^2} \cdot \frac{\sigma^2}{\sigma^2} = \frac{1}{\sigma^2}$$

Hence

$$\frac{1}{n \cdot E\left[\left(\frac{\partial \ln f(x)}{\partial \mu}\right)^2\right]} = \frac{1}{n\left(\frac{1}{\sigma^2}\right)} = \frac{\sigma^2}{n} = Var(\bar{X})$$

Hence $\bar{X}$ is a minimum variance unbiased estimator of mean (μ) of a normal distribution.

Method of Least Squares:

One aim of regression analysis is to find the smallest regression function that fits a given set of bivariate data reasonably well. Here we treat only the simplest model in which the regression function is linear.

To find the line of best fit mathematically, it is necessary to find a line which minimizes the total of the squared deviations of the actual observations from the line found. This is known as the Method of Least Squares.

Assumptions of Simple Regression

Simple regression implies that linear relationship exists between Y and X such that $Y = \alpha + \beta X$ and is based on the following assumptions:

(i) the values of X are independent

(ii) X is measured without error i.e., $E(\mu_i) = 0 \;\; \forall \; i$, where μ_i is the error term

(iii) For each value of X, there is a family (or subgroup) of values.

(iv) the variance of the sub-population are equal

(v) the means of the sub-population of Y all lie on a straight line.

Least Square Estimates

Definition: If (X_i, Y_i), $i = 1, 2, \ldots, n$ are n-points which satisfy the simple linear model, $Y = \alpha + \beta X_i + \mu_i$, the least square estimators, α and β are the values of α and β which minimize the sum of squares of the error term which is $\sum \mu_i^2$.

Derivation of Least Square Estimators

From the simple linear model:

$$Y_i = \alpha + \beta X_i + \mu_i, \quad \mu_i = Y_i - (\alpha + \beta X_i) = Y_i - \alpha - \beta X_i$$

$$S(\alpha, \beta) = \sum \mu_i^2 = \sum (Y_i - \alpha - \beta X_i)^2$$

where $S(\alpha, \beta)$ is sum of squares of the error term

$$\therefore \quad \frac{\partial S(\alpha, \beta)}{\partial \alpha} = -2 \sum (Y_i - \alpha - \beta X_i) \tag{1}$$

$$\frac{\partial S(\alpha, \beta)}{\partial \beta} = -2 X_i \sum (Y_i - \alpha - \beta X_i) \tag{2}$$

To minimize the error term,

$$\frac{\partial S(\alpha, \beta)}{\partial \alpha} = 0 \quad \text{and} \quad \frac{\partial S(\alpha, \beta)}{\partial \beta} = 0$$

Hence from (1),

$$\sum (Y_i - \alpha - \beta X_i) = 0 \tag{3}$$

and from (2),

$$X_i \sum (Y_i - \alpha - \beta X_i) = 0 \tag{4}$$

From (3),

$$\sum Y_i - \sum \alpha - \sum \beta X_i = 0$$

$$\Rightarrow \quad n\alpha = \sum Y_i - \beta \sum X_i \tag{5}$$

From (4),
$$\sum X_i Y_i = \alpha \sum X_i + \beta \sum X_i^2 \qquad (6)$$
Divide both sides of equation (5) by n
$$\therefore \hat{\alpha} = \bar{Y} - \beta \bar{X}$$

Solving equations (5) and (6) simultaneously, we obtain
$$\hat{\beta} = \frac{n \sum X_i Y_i - \sum X_i \sum Y_i}{n \sum X_i^2 - \left(\sum X_i\right)^2}$$

where $\hat{\alpha}$ and $\hat{\beta}$ are the least square estimators.

Maximum Likelihood Estimator
Definition 1

Let $f(x, \theta)$ be the probability density function of the random variable X where θ is the parameter to be estimated.

Suppose we have a sample of size n independent observations of X given by $x_1, x_2, \ldots, x_n$, the following function
$$\begin{aligned} L(X, \theta) &= f(x_1, \theta), f(x_2, \theta) \cdot f(x_3, \theta) \cdots f(x_n, \theta) \\ &= \Pi_{i=1}^{n} f_i(x_i, \theta) \end{aligned}$$
is the likelihood function of X.

Definition 2

The maximum likelihood estimator $\hat{\theta}$ of the parameter θ with the probability density function $f(x, \theta)$ is the value of θ that maximizes the likelihood function $L(X, \theta)$. Thus $\hat{\theta}$ is a function of x_i, $i = 1, 2, \ldots, n$ presumably fixed. Hence $\hat{\theta} = \hat{\theta}(x_1, x_2, \ldots, x_n)$.

Calculation of Maximum Likelihood Estimators

In calculating the maximum likelihood estimator $\hat{\theta}$ of a particular probability density function $f(x, \theta)$, we look for the value $\hat{\theta}$ which maximizes $L(X, \theta)$. To do this, we need to find

(i) $L(X, \theta)$

(ii) $\log[L(X,\theta)]$ or $\ln[L(X,\theta)]$

(iii) Obtain maximum likelihood estimator of θ by differentiating $\log[L(X,\theta)]$ partially with respect to θ and equating to zero and solve.

Example 1

A continuous random variable X with probability density function

$$f(x,\theta) = \begin{cases} \theta e^{-\theta x} &, 0 < x < \infty, \theta > 0 \\ 0 &, \text{elsewhere.} \end{cases}$$

(i) Obtain the maximum likelihood estimator of θ

(ii) If $n = 5$, $x_1 = 0.9$, $x_2 = 1.7$, $x_3 = 2.3$, $x_4 = 0.3$ and $x_5 = 0.5$. Find the value of $\hat{\theta}$

Solution

(i) $\begin{aligned} L(X,\theta) &= \Pi_{i=1}^n f(X,\theta) \\ &= \Pi_{i=1}^n \theta e^{-\theta x_i} \\ &= \theta^n e^{-\theta \Sigma x_i} \end{aligned}$

$\begin{aligned} \ln L(X,\theta) &= \ln(\theta^n e^{-\theta \Sigma x_i}) \\ &= \ln \theta^n + \ln e^{-\theta \Sigma x_i} \\ &= n \ln \theta - \theta \Sigma x_i \end{aligned}$

$\therefore \dfrac{\partial \ln L(X,\theta)}{\partial \theta} = \dfrac{n}{\theta} - \Sigma x_i$

When $\dfrac{\partial \ln L(X,\theta)}{\partial \theta} = 0$

$\dfrac{n}{\theta} - \sum x_i = 0$

$\therefore \dfrac{n}{\theta} = \sum x_i$

$\therefore \dfrac{\theta}{n} = \dfrac{1}{\sum x_i}$

$\therefore \hat{\theta} = \dfrac{n}{\sum x_i} = \dfrac{1}{\bar{X}}$

Chapter 10

$\hat{\theta} = \dfrac{1}{\bar{X}}$. Hence the maximum likelihood estimator of θ is $\dfrac{1}{\bar{X}}$

(ii) From the given information above,

$$\bar{X} = \dfrac{0.9 + 1.7 + 2.3 + 0.3 + +0.5}{5}$$

$$\therefore \hat{\theta} = \dfrac{1}{\bar{X}} = \dfrac{5}{5.7} = 0.88$$

Example 2.
Find the maximum likelihood estimator of λ in a Poisson distribution.

Solution:

$$\begin{aligned}
f(X, \lambda) &= \dfrac{e^{-\lambda}\lambda^x}{x!} \\
L(X, \lambda) &= \dfrac{e^{-\lambda}\lambda^{x_1}}{x_1!} \cdot \dfrac{e^{-\lambda}\lambda^{x_2}}{x_2!} \cdot \dfrac{e^{-\lambda}\lambda^{x_3}}{x_3!} \cdots \dfrac{e^{-\lambda}\lambda^{x_n}}{x_n!} \\
&= \dfrac{e^{-n\lambda}\lambda^{\sum x_i}}{\Pi_{i=1}^{n}x_i!}
\end{aligned}$$

$$\begin{aligned}
\log L(X, \lambda) &= -n\lambda + \sum x_i \log \lambda - \log \Pi_{i=1}^{n}x_i! \\
\dfrac{\partial L(X, \lambda)}{\partial \lambda} &= -n + \dfrac{\sum x_i}{\lambda} = 0 \\
n &= \dfrac{\sum x_i}{\lambda} \\
\Rightarrow \sum x_i &= n\lambda \\
\hat{\lambda} &= \dfrac{\sum x_i}{n} = \bar{X}
\end{aligned}$$

Hence, the maximum likelihood estimator of λ is $\bar{X}$

Example 3
Given that $y_1, y_2, \ldots, y_n$ are observations from population $N(\mu, \sigma^2)$. Find the maximum likelihood estimator of (i) μ, (ii) σ^2.

Solution

Since the observations are from $N(\mu, \sigma^2)$

$$\therefore \quad f_j(Y, \mu, \sigma^2) = \frac{1}{\sigma\sqrt{2\pi}} e^{-\frac{1}{2}\left(\frac{y_j-\mu}{\sigma}\right)^2} \tag{1}$$

$$\begin{aligned} L(Y, \mu, \sigma^2) &= \Pi_{j=1}^n f_j(Y, \mu, \sigma^2) \\ &= \frac{1}{\sigma^n (2\pi)^{n/2}} e^{-\frac{1}{2}\sum\left(\frac{y_j-\mu}{\sigma}\right)^2} \\ &= \sigma^{-n}(2\pi)^{-n/2} e^{-\frac{1}{2}\sum\left(\frac{y_j-\mu}{\sigma}\right)^2} \end{aligned} \tag{2}$$

$$\begin{aligned} \ln L(Y, \mu, \sigma^2) &= \ln \sigma^{-n}(2\pi)^{-n/2} e^{-\frac{1}{2}\sum\left(\frac{y_j-\mu}{\sigma}\right)^2} \\ &= -n\ln\sigma - \frac{n}{2}\ln 2\pi - \frac{1}{2}\sum\left(\frac{y_j-\mu}{\sigma}\right)^2 \end{aligned} \tag{3}$$

$$\therefore \quad \frac{\partial \ln L(Y, \mu, \sigma^2)}{\partial \mu} = \frac{-\frac{1}{2}(-1)(2)}{\sigma^2}\sum(y_j - \mu)$$

$$= \sum\left(\frac{y_j - \mu}{\sigma^2}\right)$$

For $\dfrac{\partial \ln L(Y, \mu, \sigma^2)}{\partial \mu} = 0,$

$$\sum\left(\frac{y_j - \mu}{\sigma^2}\right) = 0$$
$$\therefore \quad \sum y_j - n\mu = 0$$
$$\Rightarrow \quad n\mu = \sum y_j$$
$$\hat{\mu} = \frac{\sum y_j}{n} = \bar{Y}$$

Hence, the maximum likelihood estimator of μ is $\bar{Y}$.

Put $t = \sigma^2$ in (3),

$$\therefore \ln L(Y,\mu,t) = -n\ln t^{\frac{1}{2}} - \frac{n}{2}\ln 2\pi - \frac{1}{2}\frac{\sum(y_j - \mu)^2}{t}$$

$$= -\frac{n}{2}\ln t - \frac{n}{2}\ln 2\pi - \frac{1}{2}\frac{\sum(y_j - \mu)^2}{t}$$

$$\therefore \frac{\partial}{\partial t}\ln(Y,\mu,t) = -\frac{n}{2t} - 0 + \frac{\sum(y_j - \mu)^2}{2t^2}$$

$$= -\frac{n}{2t} + \frac{\sum(y_j - \mu)^2}{2t^2}$$

When $\dfrac{\partial}{\partial t}\ln(Y,\mu,t) = 0$, then

$$-\frac{n}{2t} + \frac{\sum(y_j - \mu)^2}{2t^2} = 0$$

$$\therefore \sum \frac{(y_j - \mu)^2}{2t^2} = \frac{n}{2t}$$

$$\therefore nt = \sum(y_j - \mu)^2$$

$$\therefore \hat{t} = \frac{\sum(y_j - \hat{\mu})^2}{n}$$

But $\hat{\mu} = \bar{Y}$ and $t = \sigma^2$

$$\therefore \hat{\sigma}^2 = \frac{\sum(y_j - \bar{Y})^2}{n}$$

Hence, the maximum likelihood estimator of σ^2 is

$$\hat{\sigma}^2 = \frac{\sum(y_j - \bar{Y})^2}{n}$$

ALITER

From $\ln L(Y,\mu,\sigma^2) = -n\log\sigma - \dfrac{n}{2}\ln(2\pi) - \dfrac{1}{2}\sum\left(\dfrac{y_j - \mu}{\sigma}\right)^2$

$$\therefore \frac{\partial \ln L(Y,\mu,\sigma^2)}{\partial \sigma} = \frac{-n}{\sigma} - \frac{2(-1)\sigma^{-3}}{2}\sum(y_j - \mu)^2$$

When $\dfrac{\partial \ln L(Y, \mu, \sigma^2)}{\partial \sigma} = 0$ then

$$\dfrac{-n}{\sigma} + \dfrac{\sum(y_j - \mu)^2}{\sigma^2} = 0$$
$$\therefore n\sigma^2 = \sum(y_j - \mu)^2$$
$$\therefore \hat{\sigma}^2 = \dfrac{\sum(y_j - \hat{\mu})^2}{n}$$
$$= \dfrac{\sum(y_j - \bar{Y})^2}{n}$$

Hence the maximum likelihood estimator of σ^2 is $\dfrac{\sum(y_j - \bar{Y})^2}{n}$.

Cramer Rao Lower Bound

A method of finding minimum variance unbiased estimator of a parameter is to find the Cramer-Rao lower bound for an unbiased estimator.

Definition: Let $X_1, X_2, \ldots, X_n$ be independent random variables each having density function $f(x, \theta)$, the Cramer-Rao inequality is defined as:

$$Var(\hat{\theta}) \geq \dfrac{1}{nE\left[\left(\dfrac{\partial \ln f(x)}{\partial \theta}\right)^2\right]}$$

where $\hat{\theta}$ is the unbiased estimator of θ.

Cramer Rao lower bound is given by

$$Var(\hat{\theta}) = \dfrac{1}{nE\left[\left(\dfrac{\partial \ln f(x)}{\partial \theta}\right)^2\right]}$$

Example 1.

Given that $X_1, X_2, \ldots, X_n$ is a random sample from $f(x, \lambda) = \dfrac{e^{-\lambda}\lambda^x}{x!}$ for $x = 0, 1, 2, \ldots$, obtain Cramer-Rao lower bound for the unbiased estimator λ.

Solution:

$$f(x, \lambda) = \frac{e^{-\lambda}\lambda^x}{x!}; \quad x = 0, 1, 2, \ldots$$

$$\ln f(x, \lambda) = -\lambda + x \ln \lambda - \ln x!$$

$$\frac{\partial \ln f(x)}{\partial \lambda} = -1 + \frac{x}{\lambda} - 0 = \frac{x}{\lambda} - 1.$$

$$\begin{aligned}
E\left[\frac{\partial}{\partial \lambda} \ln f(x, \lambda)\right]^2 &= E\left[\frac{x - \lambda}{\lambda}\right]^2 \\
&= \frac{1}{\lambda^2} E[x - \lambda]^2 \\
&= \frac{Var(X)}{\lambda^2}
\end{aligned}$$

But the probability distribution given above is Poisson. Hence the variance is equal to the mean i.e. $Var(X) = E(X) = \lambda$

$$\therefore E\left[\frac{\partial}{\partial \lambda} \ln f(x, \lambda)\right]^2 = \frac{Var(X)}{\lambda^2} = \frac{\lambda}{\lambda^2} = \frac{1}{\lambda}.$$

$$\therefore Var(\hat{\lambda}) \frac{1}{nE\left[\left(\frac{\partial}{\partial \lambda} \ln f(x, \lambda)\right)\right]^2} = \frac{1}{n\left(\frac{1}{\lambda}\right)} = \frac{\lambda}{n}$$

Example 2

Let $X_1, X_2, \ldots, X_n$ be a random sample from a probability distribution with probability density function

$$f(x, \lambda) = \lambda e^{-\lambda x}$$

Obtain the Cramer-Rao lower bound for the variance of an unbiased estimator for λ.

Solution

$$f(x,\lambda) = \lambda e^{-\lambda x}$$
$$\ln f(x,\lambda) = \ln \lambda - \lambda x$$
$$\frac{\partial \ln f(x,\lambda)}{\partial \lambda} = \frac{1}{\lambda} - x$$

$$\therefore E\left[\frac{\partial \ln f(x,\lambda)}{\partial \lambda}\right]^2 = E\left[\frac{1-\lambda x)}{\lambda}\right]^2$$
$$= \frac{1}{\lambda^2} E[1-\lambda X]^2$$
$$= \frac{1}{\lambda^2} Var(\lambda X)$$
$$= \frac{\lambda^2}{\lambda^2} Var(X) = Var(X)$$

The given distribution is exponential. For an exponential distribution

$$Var(X) = \frac{1}{\lambda^2}$$
$$\therefore E\left[\frac{\partial \ln f(x,\lambda)}{\partial \lambda}\right]^2 = \frac{1}{\lambda^2}$$
$$\therefore Var(\hat{\lambda}) = \frac{1}{nE\left[\frac{\partial \ln f(x,\lambda)}{\partial \lambda}\right]^2} = \frac{1}{n\left(\frac{1}{\lambda^2}\right)} = \frac{\lambda^2}{n}.$$

Interval Estimation

This is an estimator of a population parameter given by two numerical values between which the parameter may be considered to lie e.g. 99% confidence interval for the proportion (p) of non-defective batteries is $0.675 \le p \le 0.735$ is an interval estimate with lower limits 0.675 and upper limit 0.735.

Confidence interval estimation is an example of interval estimation. The wider this interval containing the population parameter is, the higher the probability that it will include the value of estimated sample statistic.

We can construct α-percent confidence level for the mean of a given example.

Definition

A $R\%$ confidence interval for some unknown population parameter θ is an interval constructed based on the results of a random sample, so that the probability that θ lies in this interval is $\dfrac{R}{100}$.

The most commonly used is 95% confidence limit. Others are 90%, 97% or 99%. If $a \leq \theta \leq b$ constitutes a 90% confidence interval for some parameter θ, we have in probability term $P(a \leq \theta \leq b) = 0.90$.

In this section we will consider interval estimation for population mean (μ), population proportion (p) and population variance (σ^2).

Confidence Interval (C.I.) for Population Mean (μ) With Known Population Variance (σ^2).

For normal distribution with mean μ and variance σ^2, $X \sim N(\mu, \sigma^2)$ and $\bar{X} \sim N\left(\mu, \dfrac{\sigma^2}{n}\right)$ where $\bar{X}$ is the mean from sample of size n.

For $\bar{X}$, the $100(1-\alpha)\%$ confidence interval for μ is given by $\bar{X} \pm Z_{\alpha/2} \dfrac{\sigma}{\sqrt{n}}$ written as $\bar{X} - Z_{\alpha/2}\sigma/\sqrt{n} \leq \mu \leq \bar{X} + Z_{\alpha/2}\sigma/\sqrt{n}$ where $\bar{X} - Z_{\alpha/2}\dfrac{\sigma}{\sqrt{n}}$ is the lower limit and $\bar{X} + Z_{\alpha/2}\dfrac{\sigma}{\sqrt{n}}$ is the upper limit.

Confidence Levels	α	$Z_{\alpha/2}$ Normal table value
90%	0.1	1.645
95%	0.05	1.96
99%	0.01	2.58

Example.
From a random sample of 144 company's employees, it was found that the average number of days each person was absent from work due to illness was 5 days per annum with a population standard deviation of 3 days. Find the confidence limits for the average number of days of absence through sickness per employee for the company as a whole.

(a) at 95% confidence level

Chapter 10

(b) at 90% confidence level.

Solution

From the question $\bar{X} = 5$, $\sigma = 3$, $n = 144$.
But $100(1 - \alpha)\%$ confidence interval for μ is given by
$$\bar{X} - Z_{\alpha/2}\frac{\sigma}{\sqrt{n}} \leq \mu \leq \bar{X} + Z_{\alpha/2}\frac{\sigma}{\sqrt{n}}$$
At 95% confidence level, $\alpha = 0.05$, $\frac{\alpha}{2} = 0.25$ and $Z_{\alpha/2} = 1.96$.

$$C.I. = 5 \pm 1.96 \left(\frac{3}{\sqrt{144}}\right)$$

or $\quad 5 - 1.96\frac{3}{\sqrt{144}} \leq \mu < 5 + 1.96\frac{3}{\sqrt{144}}$

$\therefore \quad 5 - (1.96)(\frac{1}{4}) \leq \mu < 5 + 1.96(\frac{1}{4})$

$\therefore \quad 5 - 0.49 \leq \mu \leq 5 + 0.49$

$\quad 4.51 \leq \mu \leq 5.49$

From the above the confidence limits are
lower limit = 4.51 and upper limit = 5.49.
At 90% confidence level, $\alpha = 0.1$, $Z_{\alpha/2} = 1.645$

$$C.I. = 5 \pm 1.645 \left(\frac{3}{\sqrt{144}}\right) = 5 \pm 1.645 \left(\frac{1}{4}\right) = 5 \pm 0.411$$

or $5 - 0.411 \leq \mu \leq 5 + 0.411$.
$4.589 \leq \mu \leq 5.411$.
Hence the lower limit = 4.589
and the upper limit = 5.411.

Confidence Interval for Mean μ with Unknown Variance and Small Sample

The statistic to be used in this case is $t = \dfrac{\bar{X} - \mu}{S/\sqrt{n}}$ with $(n - 1)$ degree of freedom where S is the sample standard deviation t is student t statistic and n is sample size.

Chapter 10

The $100(1-\alpha)\%$ confidence interval for μ is given by $\bar{X} \pm t_{\alpha/2}\dfrac{S}{\sqrt{n}}$ or $\bar{X} - t_{\alpha/2}\dfrac{s}{\sqrt{n}} \leq \mu \leq \bar{X} + t_{\alpha/2}\dfrac{s}{\sqrt{n}}.$

Example 1
The cost of assembling an item of computer has been estimated by obtaining a sample of 9 jobs. The average cost of assembly derived from the sample was two hundred and fifty thousand naira (₦250,000) with a standard deviation of 1.2
Construct a 90% confidence interval for the mean cost of assembling of all the items.

Solution:
The $100(1-\alpha)\%$ confidence interval for mean μ is given by $\bar{X} \pm t_{\alpha/2(n-1)}\dfrac{s}{\sqrt{n}}.$
From the question $\bar{X} = ₦250,000$, $n = 9$, $s = 1.2$
$t_{\alpha/2(n-1)} = t_{\frac{0.1}{2}(9-1)} = t_{0.05(8)} = 1.86.$
At 90% confidence level,

$$\begin{aligned} C.I. &= 250,000 \pm 1.86\dfrac{1.2}{\sqrt{9}} \\ &= 250,000 \pm 1.86\left(\dfrac{1.2}{3}\right) \\ &= 250,000 \pm 1.86(0.4) \\ &= 250,000 \pm 7.44 \end{aligned}$$

or $\quad 249999.256 \leq \mu \leq 250007.44$

Example 2
Five rabbits are measured and their lengths are found to be 2, 3, 4, 5 and 6. Find a 99% confidence limits for the mean length of a rabbit assuming these lengths form a normal distribution.
Here both mean and variance are known.

$n = 5, \quad \bar{X} = \dfrac{\sum X}{n} = \dfrac{2+3+4+5+6}{5} = 4.$

Let $d = x - \bar{X} = \tilde{x} - 4$

$$\therefore d^2 = (2-4)^2 + (3-4)^2 + (4-4)^2 + (5-4)^2 + (6-4)^2$$
$$= 4 + 1 + 0 + 1 + 4$$
$$= 10$$
$$\therefore \sum d^2 = 10$$

But
$$S^2 = \frac{\sum d^2}{n} = \frac{10}{5} = 2.$$
$$\therefore S = \sqrt{2} = 1.414.$$

From the question $n = 5$, $t_{\frac{\alpha}{2}(n-1)} = t_{0.005}(4)$.
From student t table, $t_{0.005}(4) = 4.60$.
At 99% confidence level,

$$C.I. = \bar{X} \pm t_{\frac{\alpha}{2}(n-1)} \frac{S}{\sqrt{n}} = 4 \pm 4.60 \left(\frac{1.414}{\sqrt{5}}\right)$$
$$= 4 \pm 4.60 \left(\frac{1.414}{2.236}\right)$$
$$= 4 \pm 2.91$$
$$\text{or} \quad 4 - 2.91 \leq \mu \leq 4 + 2.91$$
$$1.09 \leq \mu \leq 6.91.$$

Lower limit = 1.09, Upper limit - 6.91.

Confidence Interval For Mean μ (With Unknown Variance) and Large Sample Size

Sample size (n) is said to be large if $n \geq 25$. Then student t-statistic is approximately close to normal distribution Z.

The confidence interval for μ in this case is

$$\bar{X} \pm Z_{\alpha/2} \frac{S}{\sqrt{n}}.$$

Example
Twenty five women from the same village are taken at random and their heights

in metres are found to have a mean of 1.55 and standard deviation of 1.2. Find the 95% confidence limit of the mean height of the women.

Solution:
From the question, $n = 25$, $\bar{X} = 1.55$, $S = 1.2$
For a 95% confidence level $Z_{\alpha/2} = Z_{\frac{0.05}{2}} = Z_{0.025} = 1.96$.
$\therefore$ At 95% confidence level,

$$C.I. = \bar{X} \pm Z_{\frac{\alpha}{2}} \frac{S}{\sqrt{n}} = 1.55 \pm 1.95 \left(\frac{1.2}{\sqrt{25}}\right)$$
$$= 1.55 \pm 1.96 \left(\frac{1.2}{5}\right)$$
$$= 1.55 \pm 1.96(0.24)$$
$$= 1.55 \pm 0.4704$$
$$\text{or} \quad 1.55 - 0.4704 \leq \mu \leq 1.55 + 0.4704$$
$$1.0796 \leq \mu \leq 2.0204$$
$$\text{or} \quad 1.080 \leq \mu \leq 2.020$$

Confidence Interval for Proportion (P)

A $100(1 - \alpha)\%$ confidence interval (C.I.) for a population proportion P based on a random sample of size n is given as $\hat{p} \pm Z_2 \sqrt{\dfrac{\hat{p}q}{n}}$

$$\hat{p} - Z_{\frac{\alpha}{2}} \sqrt{\frac{\hat{p}q}{n}} < p < \hat{p} + Z_{\frac{\alpha}{2}} \sqrt{\frac{\hat{p}q}{n}}$$

where $\mu_{\hat{p}} = p$ and $Var(\sigma_p^2) = \dfrac{pq}{n}$.

Example

In a random sample of 80 graduating students who sat for an interview for an appointment, 20 of them were found appointable. Estimate the proportion of graduating students who were found appointable at 90% confidence level.

$$\hat{p} = \frac{20}{80} = \frac{1}{4} = 0.25$$
$$\hat{q} = 1 - \hat{p} = 1 - 0.25 = 0.75$$
$$Z_{\alpha/2} = Z_{\frac{0.1}{2}} = Z_{0.05} = 1.645 \text{ for 90\% C.I.}$$

$\therefore$ At 90% confidence level,

$$\text{C.I.} = 0.25 \pm 1.645\sqrt{\frac{(0.25)(0.75)}{80}} = 0.25 \pm 0.080$$

or $0.25 - 0.080 \leq p \leq 0.25 + 0.080$
$0.170 \leq p \leq 0.330$.
From the above. the lowerlimit is 0.170 and the upper limit is 0.330.

Chapter 11

Theory of Hypothesis Testing

Definitions of terms.

(1) **Simple hypothesis:** A simple hypothesis is an hypothesis which completely defines $f_j(y_j, \theta)$ e.g. $\theta = \theta_0$ where θ_0 is a numerical value.

(2) **Composite hypothesis:** This is an hypothesis in which f is not completely defined e.g. $\theta > \theta_0$ or $\theta < \theta_0$ or $\theta \neq \theta_0$.

(3) **Null hypothesis:** This is defined by $H_0 : \theta = \theta_0$. A null hypothesis nullifies the effect of a treatment and it corresponds to the absence of effects of the variable being investigated.

(4) **Alternative hypothesis:** This is denoted by $H_\wedge : \theta = \theta_\wedge$, which is defined as a range of values that would prevail if the variable being studied has some effect.

The procedure is to accept H_0 or H_A. To do this, we need (i) a test statistic T, (ii) a critical region of size $\alpha(\omega_\alpha)$ so that when T lies in ω_α, we reject H_0 and accept H_A.

Critical Region of Size $\alpha(\omega_\alpha)$: Given α, such that $0 < \alpha < 1$, ω_α is defined as the probability such that

$$P\{T \in \omega_\alpha / H_0\} = \alpha \quad \text{when } H_0 \text{ is true.}$$

Chapter 11

where T is the test statistic and ω_α is the critical region of size α. This is also the probability that you reject H_0 when H_0 is true i.e. the probability of committing type 1 error.

Best Critical Region: ω_α is called the best critical region of size α if $P(T \in \omega_\alpha / H_A)$ is maximum.

Simple null hypothesis versus simple alternatively hypothesis: This is defined by $H_0 : \theta = \theta_0$ Vs $H_A : \theta = \theta_A$.

Two Possible Errors: The two possible errors are as shown in the table below.

Table 1

	Decision	Nature	
		H_0 True	H_0 False
Statistician	Accept H_0	Correct decision	Type II error
	Reject H_0	Type I error	Correct decision

The Probability Equivalent of Table 1

If α is the probability of committing Type 1 error and β is the probability of committing Type II error.

The table below shows the probability equivalent of Table 1.

Table II

Decision	H_0 (true)	H_0 (false)
Accept H_0	$1 - \alpha$	β
Reject H_0	α	$1 - \beta$

From the two tables above, the following definitions are derived.

(1) **Type I Error:** This is the error committed when we reject H_0 when indeed H_0 is true. (H_0 is the null hypothesis).

(2) **Type II Error:** This is the error committed when we accept the null hypothesis (H_0) when in fact H_0 is false.

(3) **Power of the test:** This is the probability of rejecting the null hypothesis (H_0) when indeed it is false. It is denoted by $1 - \beta$ where β is the probability of type II error. It is also the ability of the test to accept the alternative hypothesis when it is infact true.

Summary

(1) A test must have a null hypothesis H_0 and an alternative H_A. Rejection of $H_0 \Rightarrow$ acceptance of H_A.

(2) We need a test statistic T and a critical region defined by $\omega_\alpha = \{T, P(T \in \omega_\alpha/H_\alpha) = \alpha\}$.

(3) Whenever $T \in \omega_\alpha$ we reject H_0 and we say that the test is significant.

(4) ω_α is said to be the best critical region of $P(T \in \omega_\alpha/H_A)$ is maximum.

(5) Two possible errors are Type I and Type II.
Type I error is the rejection of the null hypothesis when it is true with probability α.
Type II error is the acceptance of the null hypothesis when it is false with probability β.

(6) The power of the test $= 1 - \beta$ where β is the probability of type II error.

Likelihood Ratio Definition

Given a set of observation, $y_1, y_2, \ldots, y_n$ each from the population $f_1(y, \theta)$ and require to test $H_0 : \theta = \theta_0$ Vs $H_A : \theta = \theta_A > \theta_0$ the likelihood ratio of the required test is defined by

$$\ell_{A0} = \frac{\text{lik}/H_{\theta_A}}{\text{lik}/H_{\theta_0}} = \frac{\pi f_j(y_j, \theta_A)}{\pi f_j(y_j, \theta_0)}$$

The likelihood ratio critical region is defined by $T \in \omega_\alpha$ iff ℓ_{A0} less than or greater than C_α where C_α is chosen such that $P(\ell_{A0} \ C_\alpha/H_0) = \alpha$.

Chapter 11

Neyman Pearson Lemma

The Lemma states that: of all critical regions of size α, the likelihood ratio critical region is the best.

Example 1.

Given that $y_1, y_2, \ldots, y_n$ are observations from a population $f_j(y_j, \theta) = \dfrac{y_j e^{-y_j/\theta}}{\theta^2}$, $y_j \geq 0$. Find the best critical region of size α for testing the hypothesis $H_0 : \theta = \theta_0$ Vs $H_A : \theta = \theta_A$.

Solution

By definition, given that:

$$\ell(Y, \theta) = \prod_{j=1}^{n} \left(\frac{y_j e^{-y_j/\theta}}{\theta^2} \right)$$

$$= \frac{e^{-1/\theta \sum y_j} \prod_{j=1}^{n}(y_j)}{\theta^{2n}}$$

$$\therefore \ell_{AO} = \frac{\text{lik} H_{\theta_A}}{\text{lik}/H_{\theta_0}} = \frac{e^{-1/\theta_A \sum y_j} \prod_{j=1}^{n}(y_j/\theta_A^{2n})}{e^{-\frac{1}{\theta_0}\sum y_j} \prod_{j=1}^{n}(y_j/\theta_0^{2n})}$$

$$= e^{\sum y_j \left(\frac{1}{\theta_A} - \frac{1}{\theta_0} \right)} \times \left(\frac{\theta_0}{\theta_A} \right)^{2n} \qquad \theta_A > \theta_0$$

By Neyman Pearson Lemma, best critical region (C-R) is of form $\ell_{AO} > C_\alpha$ where

$$\ell_{AO} = e^{-\sum y_j \left(\frac{1}{\theta_A} - \frac{1}{\theta_0} \right)} \times \left(\frac{\theta_0}{\theta_A} \right)^{2n} \quad \text{i.e.} \quad \left(\frac{\theta_0}{\theta_A} \right)^{2n} e^{-\sum y_j \left(\frac{1}{\theta_A} - \frac{1}{\theta_0} \right)} > C_\alpha.$$

Taking log of both sides of the equation we obtain

$$2n(\log \theta_0 - \log \theta_A) - \left(\frac{1}{\theta_A} - \frac{1}{\theta_0} \right) \sum y_j > \log C_\alpha.$$

or

$$\left(\frac{1}{\theta_A} - \frac{1}{\theta_0} \right) \sum y_j > \log C_\alpha - 2n(\log \theta_0 - \log \theta_A)$$

Chapter 11

$$\theta_0 < \theta_A, \Rightarrow \left(\frac{1}{\theta_0} > \frac{1}{\theta_A}\right), \quad \therefore \frac{\sum y_i}{n} > \frac{\log C_\alpha - 2n(\log \theta_0 - \log \theta_A)}{n\left(\frac{1}{\theta_0} - \frac{1}{\theta_A}\right)}$$

Therefore,
$$\therefore \quad \bar{Y} > d_\alpha.$$

where d_α is the expression on the right hand side of the equation to which we can find an alternative value using the p.d.f. of $\bar{Y}$.

Now,

$$f_1(y_j, \theta) = \frac{y_j e^{-\frac{y_j}{\theta}}}{\theta^2}$$

$$= \frac{\frac{1}{\theta}\left(\frac{1}{\theta}y_j\right)^{2-1} e^{-y_j/\theta}}{\Gamma_2}$$

which is a gamma distribution with $\lambda = \frac{1}{\theta}$.

The corresponding characteristic function is

$$\phi_{y_j}(t) = \left(1 - \frac{it}{\lambda}\right)^2 = (1 - \theta it)^2 \quad \left(\lambda = \frac{1}{\theta}\right)$$

$$\phi \sum y_j(t) = (1 - \theta it)^{2n}$$

$$\phi\left(\frac{2n}{\theta}\right)\sum y_j(t) = \left(1 - \theta it \cdot \frac{2n}{\theta}\right)^{-2n} = (1 - 2nit)^{-2n}$$

$$\frac{\phi 2n \sum y_1(t)}{\theta n} = \left((1 - \frac{2nit}{n}\right)^{-2n} = (1 - 2it)^{-2n}$$

i.e. $\quad \dfrac{\phi 2n\bar{Y}(t)}{\theta} = (1 - 2it)^{-2n} = (1 - 2it)^{-\frac{4n}{2}}$

$$\Rightarrow \quad \frac{2n\bar{Y}(t)}{\theta} \text{ has a } \chi^2 \text{ distribution with } 4n \text{ degree of freedom.}$$

But the best critical region is of the form $\bar{Y} > d_\alpha$.

Let $C_{4n,\alpha}$ be chosen such that if X has a χ^2_{4n} distribution $P(X \geq C_{4n,\alpha}|H_0) = \alpha$

i.e. $\quad P\left(\dfrac{2n\bar{Y}}{\theta_0} > C_{4n,\alpha}\right) = \alpha.$

Chapter 11

$$P\left(\bar{Y} > \frac{\theta_0 C_{4n,\alpha}}{2n}\right) = \alpha$$

The best critical region of size α is $\bar{Y} > \dfrac{\theta_0 C_{4n,\alpha}}{2n}$.

Procedure for calculating the best critical region and testing the given hypothesis from a population $f_j(y_j, \theta) = \dfrac{y_j e^{-\frac{y_j}{\theta}}}{\theta^2}$, $y_j \geq 0$

(i) Calculate $\bar{Y}$

(ii) Find $C_{4n,\alpha}$ such that $P(X > C_{4n,\alpha}) = \alpha$, where X has a χ^2_{4n} distribution

(iii) Calculate $d_\alpha = \dfrac{\theta_0 C_{4n,\alpha}}{2n}$

(iv) Reject H_0 if $\bar{Y} > d_\alpha$.

Similar steps can be taken for other populations other than the one given.

Example 2.

Given that 3.28, 8.68, 10.09, 28.12 and 53.02 are the observations from the population

$$f_j(y_j, \theta) = \frac{y_j e^{-y_j \theta}}{\theta^2}, \quad y_j > 0.$$

From the above information, calculate the best critical region of size α. Use your result to test the hypothesis $H_0 : \theta = 6.0$ Vs $H_A : \theta > 6.0$ where $\alpha = 0.05$.

Solution

Since $n = 5$, the degree of freedom of χ^2 distribution $= 4n = 20$. Taking $\alpha = 0.05$, we require C such that
$P(X \geq C_{4n,\alpha}) = \alpha$ i.e. $P(X \geq C_{20, 0.05}) = \alpha$ where X has χ^2_{4n} distribution)

$$\therefore d_\alpha = \frac{\theta_0 C_{4n,\alpha}}{2n} = \frac{6 \times 31.4}{10} \quad \therefore (C_{4n,\alpha} = C_{20, 0.05} = 31.4)$$

$$= 18.846$$

$$\approx 18.85$$

Hence, the Best Critical Region of size α is $\bar{Y} > 18.85$.

$$\therefore \quad \omega_\alpha = \{\bar{Y} : \bar{Y} > 18.85\}$$

But
$$\bar{Y} = \frac{3.28 + 8.68 + 10.09 + 28.12 + 53.02}{5} = 20.64$$

Using the given data, $\bar{Y} = 20.64$.
Hence, $\bar{Y} \in \omega_\alpha$
$\therefore$ Reject H_0
Hence, $\theta > 6.0$

Example 3.
Given that $y_1, y_2, \ldots, y_n$ is a sample of observations from a population $N(\mu, \sigma_0^2)$ where σ_0^2 is known.
Test the hypothesis

$$H_0 : \mu = \mu_0 \quad \text{Vs} \quad H_A : \mu = \mu_A > \mu_0$$

Solution

$$f_j(y_j, \mu, \sigma_0) = \frac{1}{\sigma_0\sqrt{2\pi}} e^{-\frac{1}{2}\left(\frac{y_j-\mu}{\sigma_0}\right)^2}$$

$$\ell(y_j, \mu, \sigma_0) = \frac{1}{\sigma_0^n (2\pi)^{n/2}} e^{-\frac{1}{2}\sum\left(\frac{y_j-\mu}{\sigma_0}\right)^2}$$

$$\ell_{A0} = \frac{\frac{1}{\sigma_0^n (2\pi)^{n/2}} e^{-\frac{1}{2}\sum\left(\frac{y_j-\mu_A}{\sigma_0}\right)^2}}{\frac{1}{\sigma_0^n (2\pi)^{n/2}} e^{-\frac{1}{2}\sum\left(\frac{y_j-\mu_0}{\sigma_0}\right)^2}}$$

$$= e^{-\frac{1}{2}\left\{\sum\left(\frac{y_j-\mu_A}{\sigma_0}\right)^2 - \sum\left(\frac{y_j-\mu_0}{\sigma_0}\right)^2\right\}} = e^{-\frac{1}{2\sigma_0^2}\{\sum(y_j-\mu_A)^2 - \sum(y_j-\mu_0)^2\}}$$

By Neyman Pearson Lemma, best critical region is given by

Chapter 11

$\ell_{A0} > C_\alpha.$

$$\begin{aligned}
\log \ell_{A0} &= -\frac{1}{2\sigma_0^2}\left\{\sum(y_j - \mu_A)^2 - \sum(y_j - \mu_0)^2\right\} > \log C_\alpha \\
&= \frac{-1}{2\sigma_0^2}\left\{\sum(y_j - \mu_A + y_j - \mu_0)(y_j - \mu_A - y_j + \mu_0)\right\} > \log C_\alpha \\
&= \frac{1}{2\sigma_0^2}\left\{\sum[2y_j - (\mu_A + \mu_0)](\mu_A - \mu_0)\right\} > \log C_\alpha
\end{aligned}$$

$\therefore \quad \sum[2y_j - (\mu_0 + \mu_A)] > \dfrac{2\sigma_0^2 \log C_\alpha}{\mu_A - \mu_0}$

$\therefore \quad 2\sum y_j - n(\mu_0 + \mu_A) > \dfrac{2\sigma_0^2 \log C_\alpha}{\mu_A - \mu_0}$

$\sum y_j > \dfrac{1}{2}\left[\dfrac{2\sigma_0^2 \log C_\alpha}{\mu_A - \mu_0} + n(\mu_0 + \mu_A)\right]$

$\dfrac{\sum y_j}{n} > \dfrac{1}{2n}\left[\dfrac{2\sigma_0^2 \log C_\alpha}{\mu_A - \mu_0} + n(\mu_0 + \mu_A)\right]$

$\bar{Y} > \dfrac{1}{2n}\left[\dfrac{2\sigma_0^2 \log C_\alpha}{\mu_A - \mu_0} + n(\mu_0 + \mu_A)\right]$

i.e. $\bar{Y} > d_\alpha$ where d_α is the expression on the right hand side to which we want to find an alternative

$$P\{\bar{Y} > d_\alpha / H_\alpha\} = \alpha$$

Since $y_j \sim N(\mu_0, \sigma_0^2)$

$\therefore \quad Z = \dfrac{\bar{Y} - \mu}{\frac{\sigma_0}{\sqrt{n}}} \sim N(0,1)$

Here we require K_α such that

$$P\{Z \geq K_\alpha / H_0\} = \alpha$$

i.e. $P\left(\dfrac{\bar{Y} - \mu}{\dfrac{\sigma_0}{\sqrt{n}}} > K_\alpha / H_0\right) = \alpha$

$P\left(\dfrac{\bar{Y} - \mu_0}{\dfrac{\sigma_0}{\sqrt{n}}} > K_\alpha\right) = \alpha$

$\therefore \quad P\left((\bar{Y} - \mu_0) > \dfrac{K_\alpha \sigma_0}{\sqrt{n}}\right) = \alpha$

$\therefore \quad P\left(\bar{Y} > \mu_0 + \dfrac{K_\alpha \sigma_0}{\sqrt{n}}\right) = \alpha$

Test Statistic is $\bar{Y}$.

$$d_\alpha = \mu_0 + \dfrac{K_\alpha \sigma_0}{\sqrt{n}}$$

$$\omega_\alpha = \left(\bar{Y} : \bar{Y} > \mu_0 + \dfrac{K_\alpha \sigma_0}{\sqrt{n}}\right)$$

Example 4.

The following observations $105.7, 91.9, 101.8, 100.4, 89.0, 100.2, 105.5, 105.1, 107.6, 96.3$ are from normal population with variance 16 i.e. $y_j \sim N(\mu, 16)$, $j = 1, 2, \ldots, 10$.

Test $H_0 : \mu = 80$ Vs $H_A : \mu > 80$ \quad (Take $\alpha = 0.05$)

$$\bar{Y} = \dfrac{105.7 + 91.9 + 101.8 + 100.4 + 89.0 + 100.2 + 105.5 + 105.1 + 107.6 + 96.3}{10}$$

$$\bar{Y} = \dfrac{1003.5}{10} = 100.35$$

Chapter 11

Using the information above with $\alpha = 0.05$

$$K_\alpha = 1.65, \quad \frac{K_\alpha \sigma_0}{\sqrt{n}} = \frac{1.65 \times 4}{\sqrt{10}} = 2.09$$

$$\begin{aligned}
\therefore \omega_\alpha &= \left(\bar{Y} : \bar{Y} \geq \mu_0 + \frac{K_\alpha \sigma_0}{\sqrt{n}}\right) \\
&= (\bar{Y} : \bar{Y} \geq 80 + 2.09) \\
&= (\bar{Y} : \bar{Y} \geq 82.09) \\
\text{But } \bar{Y} &= 100.35 \\
\therefore \bar{Y} &\in \omega_\alpha
\end{aligned}$$

Reject H_0 and conclude that $\mu > 80$.

Example 5.

Use the same data to test

$$H_0 : \mu = 110 \quad \text{Vs} \quad H_A : \mu > 110$$

Here,

$$\mu_0 + \frac{K_\alpha \sigma_0}{\sqrt{n}} = 2.09 + 110 = 112.09$$

Hence,

$$\omega_\alpha = (\bar{Y} : \bar{Y} > 112.09)$$

But $\bar{Y} < 112.09$, hence $\bar{Y} \notin \omega_\alpha$.
We therefore accept H_0 and conclude that $\mu = 110$.

Exercises

1. For a discrete random variable X, $f(x) = \dfrac{5-x}{10}$ $x = 1, 2, 3, 4$.

    (a) Show that $f(x)$ is a probability mass function

    (b) Find (i) $E(X)$, (ii) $Var(5 - 2X)$

    Ans.: (i) 2 (ii) 4.

2. X is a random variable with probability mass function $f(x) = \left\{\dfrac{x}{10}; \quad x = 1, 2, 3, 4\right.$
    If $y = 10 - 3x$, find (i) $E(Y)$, (ii) $Var(2Y)$.

    Ans.: (i) 1 (ii) 36

3. X is a discrete random variable whose distribution is given below:

x	1	2	3	4
$f(x)$	$\frac{1}{8}$	a	b	$\frac{1}{4}$

    If $f(x)$ is a probability mass function and $a = 2b$ find the mean of X?

    Ans.: $a = \dfrac{5}{12}$, (b) $b = \dfrac{5}{24}$, mean of $X = 2\dfrac{7}{12}$

4. For a discrete random variable R, $f(r) = 5Cr/32$. What range of values of r will make $f(r)$ a probability mass function given that $0 \leq r \leq 8$?

    Ans. $r = 0, 1, 2, 3, 4, 5$

Exercises

5. X is a discrete variable with probability distribution given below:

x	0	1	2
$f(x)$	$\frac{1}{4}$	$\frac{1}{2}$	$\frac{1}{4}$

   If $\phi(x) = 2 + 3x$, find (i) $E[\phi(X)]$ (ii) $Var[\phi(X)]$.
   Ans.: (i) 5   (ii) 4.5

6. Bola threw a pair of unbiased dice, if X is a random variable that denotes the minimum number shown on the pair of dice,
   find (i) the probability distribution of X,   (ii) $E(6X - 15)$

   Ans.:(i)

x	1	2	3	4	5	6
$f(x)$	$\frac{11}{36}$	$\frac{9}{36}$	$\frac{7}{36}$	$\frac{5}{36}$	$\frac{3}{36}$	$\frac{1}{36}$

   (ii) $\frac{1}{6}$

7. A fair coin is tossed four times. Given that T is a random variable that denotes the number of heads that shows up.

   (a) Write out the sample space

   (b) Obtain the distribution of T

   (c) Use your distribution to find (i) $E(2X + 3)$

   Ans.: (a) Sample space
   $S = \{HHHH, THHH, TTHH, TTTH, HTHH,$
   $HHTH, HHHT, HHTT, HTTH, HTHT, THTH, THHT, HTTT,$
   $THTT, TTHT, TTTT\}$

   (b)

x	0	1	2	3	4
$f(x)$	$\frac{1}{16}$	$\frac{1}{4}$	$\frac{3}{8}$	$\frac{1}{4}$	$\frac{1}{16}$

   (c) (i) 7.

8. A distribution of positive integers has probability mass function $\frac{1}{31}\binom{5}{x}$ for $x = 1, 2, 3, 4, 5$.

Find (i) $31E\left(\frac{1}{2}X\right)$ (ii) $Var\left(\frac{1}{4}X\right)$.

Ans.: (i) 40 (ii) $\frac{65}{961}$

9. A writer who writes articles for a magazine finds that his proposed articles sometimes need to be revised before they are accepted for publication. The writer finds that the number of days X, spent in revising a randomly chosen article can be modeled by the following discrete possibility distribution;

x	0	1	2	4
$f(x)$	0.8	0.1	0.05	0.05

(a) Calculate (i) $E(X)$ (ii) $Var(X)$
(b) If the writer prepares a series of 15 articles for the magazine, find the expected value of the total time required for revision to these articles.
Ans.: (i) 0.4 (ii) 0.94 (b) 6 days.

10. In a gambling game three coins are tossed, a man is paid six naira if he gets all heads or all tails and he pays out one naira if either one or two heads appear.
Calculate his expected gain or loss.
Ans.: $E(X) = 0.75$ the positive sign implies gain. Hence the expected gain of the man is 0.75.

11. A random variable X with assigned probabilities is given by

$$X = \begin{cases} 0 & \text{prob } \frac{1}{8} \\ 1 & \text{prob } \frac{3}{8} \\ 2 & \text{prob } \frac{3}{8} \\ 3 & \text{prob } \frac{1}{8} \end{cases}$$

Exercises

Use the above distribution to find (i) $E(X)$, (ii) $Var\left(\frac{2}{3}X\right)$.

Ans.: (i) 1.5 (ii) $\frac{1}{3}$

12. The discrete random variable X has the probability distribution as shown below where t is a constant.

x	0	1	2	3
$f(x)$	t	$\frac{t}{2}$	$\frac{t}{4}$	$\frac{t}{20}$

Find (i) the value of t (ii) $E(9X+2)$ (iii) $Var(X)$

Ans.: (ii) $t = \frac{5}{9}$ (ii) $7\frac{3}{4}$ (iii) 0.6751.

13. A box contains two white and eight electric bulbs, Tayo draws three electric bulbs at random without replacement. The number of white balls he draws is denoted by W.

   (a) (i) Find the probability distribution of W

   (ii) Show that $P(W \leq 1) = \frac{14}{15}$

   (b) (i) Find $E(W)$ (ii) $Var(W)$

Ans.:(a)(i)

W	0	1	2
$f(W)$	$\frac{7}{18}$	$\frac{7}{15}$	$\frac{1}{15}$

(b)(i) $E(W) = \frac{3}{5}$ (ii) $Var(W) = \frac{28}{75}$

14. (a) Find the value of $E\left(\frac{X-\mu}{4\sigma}\right)^2$

   (b) Given that $Var[k(2X-3)] = Var[3(3X-5)]$, Find the value of k.

Ans.:(a) $\frac{1}{16}$ (b) $\pm\frac{9}{2}$.

15. If μ and σ^2 are the mean and the variance of the random variable X respectively and $Y = t + kx$ where t and k are constants, find (i) $E(Y)$, (ii) $Var(3Y+1)$.

Ans.: $t + k\mu$ (ii) $9k^2\sigma^2$.

Exercises 262

16. Given that $x_1, x_2, x_3, \ldots, x_n$ are independent random variables from the same distribution each of mean μ and variance σ^2. Find (i) the mean, (ii) the variance of $3x_1 + 5x_2 + 2x_3$.
    Ans.: (i) 10μ, (ii) $38\sigma^2$

17. Given that the probability density function of a random variable X is
$$f(x) = \begin{cases} \frac{1}{2}e^{-x/2} &, x \geq 0 \\ 0 &, \text{elsewhere} \end{cases}$$
    and $p(x) = e^{3x/8}$, find $E[p(x)]$.
    Ans.: 4.

18. A continuous random variable X has a p.d.f.
$$f(x) = \begin{cases} \dfrac{1}{k-t} &; t \leq x \leq k \\ 0 &; \text{elsewhere} \end{cases}$$
    (a) Identify the distribution above.
    (b) Find (i) the mean, (ii) the variance of X.
    (c) Use the result obtained in (b) to find the (i) mean, (ii) variance of a random variable X whose probability density function
$$f(x) = \begin{cases} \dfrac{1}{3} &; 1 \leq x \leq 4 \\ 0 &; \text{elsewhere} \end{cases}$$
    Ans.: (a) A rectangular distribution which lies in the interval $t \leq x \leq k$.
    (b)(i) Mean $= \dfrac{k+t}{2}$ (ii) Variance $= \dfrac{(k-t)^2}{12}$
    (c) (i) Mean $= 2.5$ Variance $= 0.75$.

19. A continuous random variable X has the p.d.f.
$$f(x) = \begin{cases} px^2 &; 0 < x < 2 \\ px &; 0 < x < 4 \\ 0 &; \text{elsewhere} \end{cases}$$

Exercises

Find (i) the constant p, (ii) mean of X.
Ans.: $p = \frac{3}{32}$ (ii) $2\frac{3}{8}$

20. A continuous random X has the p.d.f.

$$f(x) = \begin{cases} pe^{-x/5} & ; x \geq 0 \\ 0 & ; \text{elsewhere} \end{cases}$$

Find (i) the constant p, (ii) $E(e^{x/6})$.
Ans.: $p = \frac{1}{5}$ (ii) 6

21. A continuous random variable X has a probability density function p.d.f.
$$f(x) = \begin{cases} ke^{-3x}, & x > 0 \\ 0, & \text{elsewhere} \end{cases}$$
Find (i) the constant k, (ii) $P(x < 0)$
Ans.: (i) $k = 3$, (ii) 0

22(a) What value of d makes the function

$$f(x) = \begin{cases} d(2-x), & 0 < x < 2 \\ 0, & \text{elsewhere} \end{cases}$$

a probability density function?

(b) Use your result to find the following
(i) $P(1 < x < 2)$ (ii) $P(3 < x < 5)$
Ans.: (a) $d = \frac{1}{2}$ (b)(i) 0.25 (ii) 0

23. A random variable X has a p.d.f

$$f(x) = \begin{cases} 1 - \dfrac{dx}{6}, & 0 < x < 3 \\ 0, & \text{elsewhere} \end{cases}$$

Find (i) the constant d (i) $P(-3 < x < 2)$ (iii) $P(1 < x < 4)$
Ans.: (i) $\dfrac{8}{3}$ (ii) $\dfrac{10}{9}$ (iii) $\dfrac{2}{9}$

Exercises

24. X and Y are two continuous random variable with joint p.d.f.

$$f(x,y) = \begin{cases} rx^2y &, \ 0 < x < 2, \ 1 \leq y \leq 3 \\ 0 &, \ \text{elsewhere} \end{cases}$$

Find (i) r (ii) $E(X)$ (iii) $E(Y)$

Ans.: (i) $\dfrac{3}{32}$ (ii) $1\dfrac{1}{2}$ (iii) $2\dfrac{1}{6}$

25. Given that X and Y are two continuous random variables with joint density function,

$$f(x,y) = \begin{cases} t(2x + 3y) &, \ 0 < x < 1, \ 0 \leq y \leq 2 \\ 0 &, \ \text{elsewhere} \end{cases}$$

Find (i) the constant t; (ii) $P(0 < X < 2, \ 0 \leq Y \leq 5)$

Ans.: $\dfrac{1}{8}$ (ii) 1.0

26. X and Y are two continuous random variables with joint p.d.f.

$$f(x,y) = \begin{cases} d(x + y) &; \ 0 \leq x \leq 2, \ 0 \leq y \leq 1 \\ 0 &, \ \text{elsewhere} \end{cases}$$

Find (i) the constant d, (ii) $E(X)$, (iii) $Var(X)$

Ans.: (i) $\dfrac{1}{3}$ (ii) $\dfrac{11}{9}$ (iii) $\dfrac{23}{81}$

27. The joint probability density function of two continuous random variables is given by

$$f(x,y) = \begin{cases} q(2x^2 + y^2) &; \ 0 \leq x \leq 2, \ 0 \leq y \leq 3 \\ 0 &, \ \text{elsewhere} \end{cases}$$

Find (i) q (ii) $P(1 \leq X \leq 2, \ Y \geq 1)$ (iii) $E(Y)$.

Ans.: (i) $\dfrac{1}{34}$ (ii) $\dfrac{9}{17}$ (iii) $E(Y) = \dfrac{124}{68}$

28. The random variables X_1, X_2 have joint function

$$f(x_1, x_2) = \begin{cases} 4x_1x_2 &; \ 0 \leq x_1 \leq 1, \ 0 < x_2 < 1 \\ 0 &, \ \text{elsewhere} \end{cases}$$

Exercises

(i) Show that $f(x_1, x_2)$ is a probability density function

(ii) Find $P(0 < X_1 < \frac{1}{2}, \frac{1}{4} < X_2 < 1)$

Ans.: (ii) $\dfrac{15}{64}$

29. X_1, X_2 are random variables having joint probability density function

$$f(x_1, x_2) = \begin{cases} k(x_1 + 3x_2) & ; \ 0 < x_1 < 1, \ 0 < x_2 < 2 \\ 0 & , \ \text{elsewhere} \end{cases}$$

Find (i) the constant k, (ii) marginal density function of X_1, (iii) the variance of X_1 given X_2

Ans.: (i) $\dfrac{1}{7}$ (ii) $\dfrac{2x_1 + 3}{7}$ (iii) $\dfrac{54x_2^2 + 18x_2 + 1}{18(1 + 6x_3)^2}$

30. If X and Y are random variables with joint density function

$$f(x, y)) = \begin{cases} r(3 - x + y) & ; 0 < x < 2, \ 0 < y < 2 \\ 0 & ; \ \text{elsewhere} \end{cases}$$

find (i) the constant r, (ii) the marginal probability density function of Y

(iii) the conditional probability density function of X given Y.

(iv) the conditional expectation of X given Y

Ans.: (i) $r = \dfrac{1}{12}$ (ii) $\dfrac{2 + y}{6}$ (iii) $\dfrac{3 - x + y}{2(2 + y)}$ (iv) $\dfrac{5 + 3y}{3(2 + y)}$

31. (a) What value of p makes the function

$$f(x) = \begin{cases} p(x + 2) & ; 0 < x < 2 \\ 0 & ; \text{elsewhere} \end{cases}$$

a probability density function?

(b) Use your result to find the following

(i) $P(1 < X < 2)$ (ii) $P(3 < X < 7)$

(c) Find (i) $E(3X+1)$ (ii) $Var(3X+1)$.

Ans.: (a) $\dfrac{1}{6}$

(b(i)) $\dfrac{7}{12}$ (ii) 0

(c) (i) $\dfrac{13}{3}$ (ii) $\dfrac{26}{9}$

Binomial Distribution

32. Five fair coins are tossed, find the probability of obtaining:

    (i) three heads

    (ii) at least three heads

    (iii) at most three heads

    Ans.: (i) $\dfrac{5}{16}$ (ii) $\dfrac{1}{2}$ (iii) $\dfrac{13}{16}$

33. Two parents (husband and wife) are tall. For each of the children, the probability of being tall is $\dfrac{1}{4}$. If they have four children, find the probability that exactly two of them will be tall.
    Ans.: $\dfrac{27}{128}$.

34. It is known that 30% of the adult population in a particular University in Nigeria registered with the University Health Center for immunisation. If a random sample of 10 adults are gathered in the institution sports complex, what is the probability that there are
    (i) at most 2
    (ii) at leat 4
    registered adults in the Health Centre are in the gathering.
    Ans.: (i) 0.3828   (ii) 0.3504

35. If the birth of a baby boy and a baby girl are equiprobable, find the probability that in a family of five children:

    (i) exactly 3 will be boys

Exercises

(ii) at most one will be a boy

Ans.: (i) $\dfrac{5}{16}$ (ii) $\dfrac{3}{16}$

36. A fair die is tossed three times. Find the probability that either a multiple of 2 or a multiple of 3 appears:

    (i) exactly one time

    (ii) more than two times

    Ans.: (i) $\dfrac{5}{72}$ (ii) $\dfrac{125}{216}$

37. A gardener makes a gain of five dollars on the average per day on the sale of a particular commodity in his garden. What is the probability that on a particular day he will make
    (i) no gain
    (ii) less than two dollars
    (iii) at least two dollars
    (iv) less than three dollars.
    Ans.: (i) 0.00698 (ii) 0.04188 (iii) 0.9581 (iv) 0.1291

38. Given that $n = 400$, $p = 0.01$, find (i) $P(X \geq 3)$ (ii) $P(3 \leq X < 6)$
    Ans.: 0.7618 (ii) 0.5470.

39. The following tables show the number of customers who deposited over a half a million naira with a commercial bank for 50 days period.

Number of customers	0	1	2	3	4
Number of days	13	15	12	7	3

    (a) Fit a poisson distribution to the data.

    (b) Use the distribution fitted to find $P(X < 2)$.
    Ans.: (a) $P(X = x) = \dfrac{e^{-1.44} 1.44^x}{x!}$ $x = 0, 1, 2, 3, 4$
    (b) 0.5780

Exercises 268

40. An insurance company is considering starting a policy for HIV patients which affects 0.2% of the population. If a random sample of 2000 individuals is selected:

    (a) find the probability that at most six and at least three will have HIV.

    (b) If the company is not willing to introduce the policy if more than 2 out of 2000 have HIV, calculate the probability that the policy will be introduced.

    Ans.: (i) 0.6512   (b) 0.2381

41. Two out of two thousand people reacted to a newly manufactured vaccines against tuberculosis. If 3000 people were treated with this vaccine, find the probability that:

    (i) exactly 2people reacted to the vaccine

    (ii) at most 2 people reacted to the vaccine

    (iii) at least 3 people reacted to the vaccine

    Ans.: (i) 0.224   (ii) 0.423   (iii) 0.577

42. The probability that a person gets a reaction from a new drug in the market is 0.001. If 2000 people are treated with this drug, find approximately, the probability that:

    (i) exactly three people will get a reaction

    (ii) more than two people will get a reaction

    Ans.: (i) 0.1804   (ii) 0.3233

43. Four percent of all those who were attacked by a particular disease eventually die. Eighty people attacked by this disease were randomly selected.
    Find the probability that;

Exercises

(i) at most two will die
(ii) between two and four inclusive will die.
Ans.: (i) 0.3799 (ii) 0.6044.

44. $X \sim N(240, 625)$, find:
    (i) $P(X < 230)$ (ii) $P(250 < X < 280)$ (iii) $P(X > 300)$
    Ans.: 0.3446 (ii) 0.2898 (iii) 0.0082

45. Find the value of t in the following cases:
    (i) $P(Z < t) = 0.9115$
    (ii) $P(Z \geq t) = 0.0885$ (iii) $P(|Z| < t) = 0.6046$
    Ans.: (i) 1.35 (ii) 1.35 (iii) 0.85

46. The mean and standard deviation in a statistics quiz are 60 and 15 respectively, find the score in standard units of the students receiving
    (i) 15 marks (ii) 90 marks
    Ans.: (i) -3 (ii) 2

47. Use the conditions given in quesstion 46, find the marks corresponding to the standard score
    (i) 0 (ii) 1.50
    Ans.: (i) 60 (ii) 82.5

48. The score of students in a Physics examination are normally distributed with mean 55 percent and standard deviation 10 percent. Find the probability that a particular student scores (i) less than 75 percent, (ii) between 60 and 70 percent.
    Ans.: (i) 0.9772 (ii) 0.2417

49. The quality grade-point averages of 500 college fresh men follow approximately a normal distribution with mean 2.0 and a standard deviation 1.0.

    (a) What percentage of these fresh men had between 2.5 and 3.5 grade-point average?

    (b) How many of them (to the nearest whole number) fell into the category (a) above?

Exercises   270

Ans.: (a) 24.17%   (b) 121.

50. The mean weight of 700 students are normally distributed with mean 62kg and standard deviation of 4 kg. How many students to the nearest whole number have weights:
    (i) above 70 kg
    (ii) less than 65 kg
    Ans.: (i) 16   (ii) 541.

51. If 0.02% of cocoa seedlings for exportation are infected, find the probability that out of a sample of 5000 seedlings:
    (i) more than 15 are infected
    (ii) between 12 and 18 are infected.
    Ans.: (i) 0.0571   (ii) 0.2583

52. Given that X is normally distributed with mean 25 and variance 16. Find $P(X < 23)$ for a random sample of 16.
    Ans.: 0.0228

53. The mean number of accident per day in a given city is 10 with a standard deviation of 2. Compute the probability that over a period of 16 days an observer will record a mean accident rate of:
    (i) 11 per day or less
    (ii) between 9 and 11 per day. (Give answer to two decimal places)
    Ans.: (i) 0.98   (ii) 0.96.

54. The monthly earnings of a population of workers in a particular Oil Company are normally distributed with mean 60 dollars and a standard deviation 25 dollars. Find the probability that a sample of 100 workers drawn from this population will have a weekly earning of
    (i) less than 57
    (ii) between 55 and 65
    Ans.: 0.1151   (ii() 0.9544

55. The discrete random variable X has probability distribution $P(X = 0) = \frac{1}{5}, P(X = 1) = \frac{3}{10}, P(X = 2) = \frac{2}{5}, P(X = 3) = \frac{1}{10}$.

Exercises

Find the probability that a random sample of 100 observations on X will have a total more than 150.
Ans.: 0.1379

56. The probability of having a male graduate in a family is 0.6. Find the probability that family's fifth child is the second graduate.
Ans.: 0.0922

57. There is a vacancy for the post of a Chief Accountant in a company. Seven people applied. Three of them have degrees in Accounting. If four of these applicants are randomly chosen for an interview, what is the probability that two of them have degrees in Accounting?
Ans.: $\dfrac{18}{35}$.

58. The probability that a candidate will pass a competitive examination in Mathematics on any given trial is 0.4. Find the probability that a candidate will finally pass the examination the fourth trial.
Ans.: 0.0864.

59. The probability that a child exposed to a certain contagious disease will catch it is 0.40. What is the probability that the tenth child exposed to the disease will be the third to catch it.
Ans.: 0.0645

60. A clinic has twelve health workers, five of them are nurses, four are clinic assistants, three are medical doctors, find the probability that out of six health workers selected, three are nurses, two are clinic assistants and one is a medical doctor.
Ans.: 0.121.

61. The discrete random variable X has probability distribution
$P(X = 0) = \dfrac{1}{4},\ P(X = 1) = \dfrac{1}{2},\ P(X = 2) = \dfrac{1}{4},$

   (a) find the first four moments about the origin

   (b) Use your result to find the second and third moment abaout the mean.

Exercises

Ans.: (a) $\mu'_1 = 1$, $\mu'_2 = \dfrac{3}{2}$, $\mu'_3 = \dfrac{5}{2}$, $\mu'_4 = \dfrac{9}{2}$

(b)(i) $\mu_3 = \dfrac{1}{2}$, $\mu_3 = 0$

62. X is a discrete random variable with probability mass function
    $f(x) = k(2x - 1)$, $x = 1, 2, 3$, (k is a constant)

   (a) Find the value of k.

   (b) Find the first three moments about the origin

   (c) Use your result to find (i) the variance of X, (ii) the third factorial moment about the origin.

   Ans.: (a) $k = \dfrac{1}{9}$ (b) $\mu'_1 = \dfrac{22}{9}$ $\mu'_2 = \dfrac{58}{9}$, $\mu'_3 = \dfrac{160}{9}$ (c)(i) $\dfrac{38}{81}$ (ii) $\dfrac{10}{3}$

63. A continuous random variable X has probability density function

$$f(x) = \begin{cases} kx & 0 \le x \le 1 \\ k - x & 1 \le x \le 2 \\ 0 & \text{elsewhere} \end{cases} \quad (k \text{ is a constant})$$

   (a) Find the value of k

   (b) Find the first and second moments about the origin

   (c) Use your result to find the second factorial moment about the origin.

   Ans.: (a) $k = \dfrac{5}{3}$ (b) $\mu'_1 = \dfrac{13}{18}$, $\mu'_2 = \dfrac{5}{9}$ (c) $\mu'_{[2]} = \dfrac{-1}{6}$

64. (a) Show that for a rectangular distribution of unit interval, the rth moment about the origin is $\dfrac{1}{r+1}$

   (b) Use your result to find (i) the second and fourth moment about the mean (ii) the third factorial moment about the origin.

   Ans.: (b)(i) $\mu_2 = \dfrac{1}{12}$, $\mu_4 = \dfrac{1}{80}$, (ii) $\mu'_{[3]} = \dfrac{1}{4}$

Exercises
273

65. X is a random variable with probability density function $f(x) = e^{-x}$, $0 \leq x \leq \infty$,
find (a) the mean (b) the variance of Y if (i) $y = 5x$ (ii) $y = \frac{1}{2}x$.

Ans.: (i) When $y = 5x$, $E(Y) = 5$, $Var(Y) = 25$ (ii) When $y = \frac{1}{2}x$,
$E(Y) = \frac{1}{2}$, $Var(Y) = \frac{1}{4}$

66. X is a continuous random variable with probability density function
$$f(x) = \begin{cases} xe^{-x} &, x \geq 0 \\ 0 &, \text{elsewhere} \end{cases}$$
find (i) the second (ii) third factorial moment of X about the origin.
Ans.: (i) 4 (ii) 10

67. The probability density function of a random variable
$$f(x) = \begin{cases} k &, 0 < x < 2 \\ 0 &, \text{elsewhere} \end{cases} \quad (k \text{ is a constant})$$

(a) Find the value of k.
(b) Show that the rth moment about the origin of the distribution is $\dfrac{2^r}{r+1}$
(c) Use your result to find (i) the first three moments about the origin (ii) $Var(X)$, (iii) third factorial moment about the origin.
Ans.: (i) $k = \dfrac{1}{2}$ (c) $\mu_1' = 1$, $\mu_2' = \dfrac{4}{3}$, $\mu_3' = 2$ (ii) $Var(X) = \dfrac{1}{3}$,
(iii) $\mu_{[3]}' = 0$

68. Given that the moment generating function of a random variable X with probability density function
$$f(x) = \begin{cases} k(kx)^{n-1} e^{-kx} &; x \geq 0 \\ 0 &; \text{elsewhere} \end{cases}$$
is $M_X(t) = \left(1 - \dfrac{t}{k}\right)^{-n}$

(a) Use the result above to find the first, second and third moments about the origin of a random variable X with probability density function

$$f(x) = \begin{cases} \dfrac{3(3x)^5 \, e^{-3x}}{\Gamma_2} & ; x > 0 \\ 0 & ; \text{elsewhere} \end{cases}$$

(b) Find the second factorial moment about the origin.

Ans.: (a) $\mu'_1 = 2$, $\mu'_2 = \dfrac{14}{3}$, $\mu'_3 = \dfrac{112}{9}$ (b) $\dfrac{8}{3}$

69. (a) Find the moment generating function of a random variable X with probability density function

$$f(x) = \begin{cases} 3e^{-3x} & ; x \geq 0 \\ 0 & ; \text{elsewhere} \end{cases}$$

(b) From the moment generating function obtained, find the positive difference between the 2nd moment about the mean and the third factorial moment about the origin:

Ans.: (i) $\left(1 - \dfrac{t}{3}\right)^{-1}$ (b) $\dfrac{1}{9}$

70. A random variable X has probability density function

$$f(x) = \begin{cases} (d-c)e^{-x(d-c)} & ; x \geq 0 \\ 0 & ; \text{elsewhere} \end{cases}$$

(a) Find the moment generating function of X with probability density function given above.

(b) Use your result to find the first three moments about the origin.

(c) Hence or otherwise find the mean and variance of a random variable Y whose probability density function is

$$f(x) = \begin{cases} \dfrac{1}{5} e^{-\frac{y}{5}} & ; y > 0 \\ 0 & ; \text{elsewhere} \end{cases}$$

Exercises 275

Ans.: (a) $\left(1 - \dfrac{t}{d-c}\right)^{-1}$

(b) $\mu'_1 = \dfrac{1}{d-c}$, $\mu'_2 = \dfrac{2}{(d-c)^2}$, $\mu'_3 = \dfrac{6}{(d-c)^3}$

(c) $E(Y) = 5; Var(Y) = 25.$

71. A random variable Y has a moment generating function
    $M(t) = \exp 3(e^{3t} + 5)$
    (a) Find the commulant generating function of $M(t)$
    (b) Hence find the mean and the variance of Y.
    Ans.: (a) $3(e^{3t} + 5)$ (b) $E(Y) = 9$ $Var(Y) = 27$

72. Given that the moment generating function of a distribution is $(1-2t)^{-\frac{k}{2}}$,
    (a) Show that its commulant generating function is $\dfrac{-k}{2} \ln(1 - 2t)$ and its variance is $2k$.

73. A random variable X has a moment generating function $M(t) = \left[\dfrac{1}{2}(1 + e^t)\right]^2$
    use it to evaluate (i) the mean and (ii) the variance of X.
    Ans.: $E(X) = 1$ $Var(X) = \dfrac{1}{2}.$

74. Show that $S^2 = \dfrac{1}{n}\sum_{x=1}^{n}(X_i - \bar{X})^2$ is a biased estimator of population variance σ^2.

75. Given that $x_1, x_2, \ldots, x_n$ is a random sample from a distribution with probability density function
    $f(X;\theta) = \theta^x(1-\theta)^{1-x}, \quad x = 0, 1, \quad 0 < \theta < 1.$
    Show that $\sum_{i=1}^{n} x_i$ is a sufficient statistics for θ.

76. A random variable X has a probability density function $f(x, \alpha) = \dfrac{e^{-\alpha t}(\alpha t)^x}{x!}.$
    Find the maximum likelihood estimator of α.

Exercises

Ans.: $\hat{\alpha} = \dfrac{\bar{x}}{t}$.

77. Given that the probability density function of a random variable Y is $f(y, \theta) = \theta(1 - \theta)^{y-1}$, Find the maximum likelihood estimator of θ.
Ans.: $\hat{\theta} = \dfrac{1}{\bar{y}}$.

78. A random variable X is $N(0, 1)$.
Given that $Y = \alpha + \beta x + \delta x^2$, show that
(i) $E(Y) = \alpha + \delta$

79. Given that $X \sim N(0, 1)$. Show that the moment generating function of $y = x^2$ is $\dfrac{1}{\sqrt{1 - 2t}}$.

80. A random variable T has $E(T) = 6 \quad Var(T) = 9$. Using Chebychev's inequality, find (i) the value of k for which $P(|x - 6| \geq 3k) \leq \dfrac{1}{16}$
(ii) the value of c for which $P(|X - 6| \leq c) \geq 0.96$.
Ans.: (i) 4   (ii) 15.

81. A random variable X has $\mu = 18$ and variance 16. Using Chebychev inequality find $P(|X - 18| > 6)$.
Ans.: $P(|X - 18| > 6) \leq \dfrac{4}{9}$

82. A random variable X has $\mu = 10$ and $\sigma^2 = 0.04$. Use Chebychev inequality to find $P(7 < X < 13)$
Ans.: $P(|X - 10| < 3) \geq \dfrac{224}{225}$.

83. (a) The number of bottles produced in a bottling company is a random variable with mean 300 and unknown variance. Find the probability that in a given week, the production will be at least 2000.
And.: $\dfrac{3}{20}$

Exercises 277

84. An examination consists of 100 objective questions. Five possible answers are provided for each question. A candidate without previous knowledge of the subject is given this question to be attempted by guess work. If he scores 30% or more, then he has passed the examination.

   (a) What is the probability that this candidate will pass by more guess work?

   (b) If the examination is given to 2000 such candidates, how many of them will pass the exam?

   Ans.: (a) 0.0062   (b) 12.

85. A market researcher performs a survey in order to determine the popularity. Omo washing detergent in Lagos metropolis. He visits every housing estate in Lagos and ask the question:
   "Do you use Omo detergent? Of 235 people questioned 75 answered Yes!. Treating the sample as being random, calculate a 95% confidence interval for the proportion of households in Lagos which use Omo detergent.
   Ans.: $0.2596 < p < 0.3788$

86. A container has twelve items of which five are defective, four items are drawn at random from the container one after the other without replacement. Find the probability that all the four are non-defective. **Ans.:** $\dfrac{7}{99}$

87. Ten fish of the same specie are measured and their length in centimeters are respectively found to be 12, 8, 1, 5, 3, 7, 5, 6, 9, 14.
   Find a 95% confidence interval for the mean length of a fish.
   Ans.: $1.259 < \mu < 18.741$

88. The marks in Statistics of a class are normally distributed with mean μ and standard deviation 4. A sample of 9 students from this class gave a mean mark of 60.
   Test the hypothesis $Ho : \mu = 50$ Vs $H_1 : \mu > 50$ at 5% level of significance.
   Ans.: We reject H_0 and conclude that $\mu > 50$.

Exercises   278

89. A managing director believes that the mean daily income of messengers in his bottling company is ₦1,000. Suppose a random sample of 144 workers is taken and a mean daily income of ₦1,200 found. If the population standard deviation is known to be ₦300. Check the claim of the managing director based on the sample information at $\alpha = 0.10$ level of significance.
    Ans.: We reject the null hypothesis and conclude that $\mu \neq$ ₦1,000.

90. A random sample of six observations taking from a population with mean μ and unknown variance σ^2 gives the following observations 6, 4, 10, 2, 8, 5. Test the hypothesis $Ho : \mu = 6.5$ Vs $H_1 : \mu < 6.5$ at 5% level of significance.
    Ans.: We reject H_1 and conclude that the mean $\mu = 6.5$.

Tables

Area Under Standard Normal Curve

z	0	1	2	3	4	5	6	7	8	9
0.0	.0000	.0040	.0080	.0120	.0160	.0199	.0239	.0279	.0319	.0359
0.1	.0398	.0438	.0478	.0517	.0557	.0596	.0636	.0675	.0714	.0753
0.2	.0793	.0832	.0871	.0910	.0948	.0987	.1026	.1064	.1103	.1141
0.3	.1179	.1217	.1255	.1293	.0331	.1368	.1406	.1443	.1480	.1517
0.4	.1554	.1519	.1628	.1664	.1700	.1736	.1772	.1808	.1844	.1870
0.5	.1915	.1950	.1985	.2019	.2054	.2088	.2123	.2157	.2190	.2224
0.6	.2257	.2291	.2324	.2357	.2389	.2422	.2454	.2486	.2517	.2549
0.7	.2580	.2611	.2642	.2673	.2704	.2734	.2764	.2794	.2823	.2852
0.8	.2881	.2910	.2939	.2967	.2995	.3023	.3051	.3078	.3106	.3133
0.9	.3159	.3186	.3212	.3238	.3264	.3289	.3315	.3340	.3365	.3389
1.0	.3413	.3438	.3461	.3485	.3508	.3531	.3554	.3577	.3599	.3621
1.1	.3643	.3665	.3686	.3708	.3729	.3749	.3770	.3790	.3810	.3830
1.2	.3849	.3869	.3888	.3907	.3925	.3944	.3962	.3980	.3997	.4015
1.3	.4032	.4049	.4066	.4082	.4099	.4115	.4131	.4147	.4162	.4177
1.4	.4192	.4207	.4222	.4236	.4251	.4265	.4279	.4292	.4306	.4319
1.5	.4332	.4345	.4357	.4370	.4382	.4394	.4406	.4418	.4429	.4441
1.6	.4452	.4463	.4474	.4484	.4495	.4505	.4515	.4525	.4535	.4545
1.7	.4554	.4564	.4573	.4582	.4591	.4599	.4608	.4616	.4625	.4633
1.8	.4641	.4649	.4656	.4664	.4671	.4678	.4686	.4693	.4699	.4706
1.9	.4713	.4719	.4726	.4732	.4738	.4744	.4750	.4756	.4761	.4767
2.0	.4772	.4778	.4783	.4788	.4793	.4798	.4803	.4808	.4812	.4817
2.1	.4821	.4826	.4830	.4834	.4838	.4842	.4846	.4850	.4854	.4857
2.2	.4861	.4864	.4868	.4871	.4875	.4878	.4881	.4884	.4887	.4890
2.3	.4893	.4806	.4898	.4901	.4904	.4906	.4909	.4911	.4913	.4916
2.4	.4918	.4920	.4922	.4925	.4927	.4929	.4931	.4932	.4934	.4936
2.5	.4938	.4940	.4941	.4943	.4945	.4946	.4948	.4949	.4951	.4952
2.6	.4953	.4955	.4956	.4957	.4959	.4960	.4961	.4962	.4963	.4964
2.7	.4965	.4966	.4967	.4968	.4969	.4970	.4971	.4972	.4973	.4974
2.8	.4974	.4975	.4976	.4977	.4977	.4978	.4979	.4979	.4980	.4981
2.9	.4981	.4982	.4982	.4983	.4984	.4984	.4985	.4985	.4986	.4986
3.0	.4987	.4987	.4987	.4988	.4988	.4989	.4989	.4989	.4990	.4990
3.1	.4990	.4991	.4991	.4991	.4992	.4992	.4992	.4992	.4993	.4993
3.2	.4993	.4993	.4994	.4994	.4994	.4994	.4994	.4995	.4995	.4995
3.3	.4995	.4995	.4995	.4996	.4996	.4996	.4996	.4996	.4996	.4997
3.4	.4997	.4997	.4997	.4997	.4997	.4997	.4997	.4997	.4997	.4998
3.5	.4998	.4998	.4998	.4998	.4998	.4998	.4998	.4998	.4998	.4998
3.6	.4998	.4998	.4999	.4999	.4999	.4999	.4999	.4999	.4999	.4999
3.7	.4999	.4999	.4999	.4999	.4999	.4999	.4999	.4999	.4999	.4999
3.8	.4999	.4999	.4999	.4999	.4999	.4999	.4999	.4999	.4999	.4999
3.9	.5000	.5000	.5000	.5000	.5000	.5000	.5000	.5000	.5000	.5000

Percentile Values (t_p) for Student's Distribution with ν Degree of Freedom (shaded area = p)

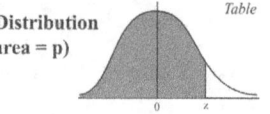

ν	$t_{.995}$	$t_{.99}$	$t_{.975}$	$t_{.95}$	$t_{.90}$	$t_{.80}$	$t_{.75}$	$t_{.70}$	$t_{.60}$	$t_{.55}$
1	63.66	31.82	12.71	6.31	3.08	1.376	1.000	.727	.325	.158
2	9.92	6.96	4.30	2.92	1.89	1.061	.816	.617	.289	.142
3	5.84	4.54	3.18	2.35	1.64	.978	.765	.584	.277	.137
4	4.60	3.75	2.78	2.13	1.53	.941	.741	.569	.271	.134
5	4.03	3.36	2.57	2.02	1.48	.920	.727	.559	.267	.132
6	3.71	3.14	2.45	1.94	1.44	.906	.718	.553	.265	.131
7	3.50	3.00	2.36	1.90	1.42	.896	.711	.549	.263	.130
8	3.36	2.90	2.31	1.86	1.40	.889	.706	.546	.262	.130
9	3.25	2.82	2.26	1.83	1.38	.883	.703	.543	.261	.129
10	3.17	2.76	2.23	1.81	1.37	.879	.700	.542	.260	.129
11	3.11	2.72	2.20	1.80	1.36	.876	.697	.540	.260	.129
12	3.06	2.68	2.18	1.78	1.36	.873	.695	.539	.259	.128
13	3.01	2.65	2.16	1.77	1.35	.870	.694	.538	.259	.128
14	2.98	2.62	2.14	1.76	1.34	.868	.692	.537	.258	.128
15	2.95	2.60	2.13	1.75	1.34	.866	.691	.536	.258	.128
16	2.92	2.58	2.12	1.75	1.34	.865	.690	.535	.258	.128
17	2.90	2.57	2.11	1.74	1.33	.863	.689	.534	.257	.128
18	2.88	2.55	2.10	1.73	1.33	.862	.688	.534	.257	.127
19	2.86	2.54	2.09	1.73	1.33	.861	.688	.533	.257	.127
20	2.84	2.53	2.09	1.72	1.32	.860	.687	.533	.257	.127
21	2.83	2.52	2.08	1.72	1.32	.859	.686	.532	.257	.127
22	2.82	2.51	2.07	1.72	1.32	.858	.686	.532	.256	.127
23	2.81	2.50	2.07	1.71	1.32	.858	.685	.532	.256	.127
24	2.80	2.49	2.06	1.71	1.32	.857	.685	.531	.256	.127
25	2.79	2.48	2.06	1.71	1.32	.856	.684	.531	.256	.127
26	2.78	2.48	2.06	1.71	1.32	.856	.684	.531	.256	.127
27	2.77	2.47	2.05	1.70	1.31	.855	.684	.531	.256	.127
28	2.76	2.47	2.05	1.70	1.31	.855	.683	.530	.256	.127
29	2.76	2.46	2.04	1.70	1.31	.854	.683	.530	.256	.127
30	2.75	2.46	2.04	1.70	1.31	.854	.683	.530	.256	.127
40	2.70	2.42	2.02	1.68	1.30	.851	.681	.529	.255	.126
60	2.66	2.39	2.00	1.67	1.30	.848	.679	.527	.254	.126
120	1.26	2.36	1.98	1.66	1.29	.845	.677	.526	.254	.126
∞	2.58	2.33	1.96	1.645	1.28	.842	.674	.524	.253	.126

Source: R.A Fisher and F. Yates, Statistical Tables for Biological, Agricultural and Medical Research (5th Edition). Table 111, Oliver and Boyd Ltd, Edinburgh, by permission of the authors and publishers, called from 'Statistical" 3rd Edition. Shatum's Outline Series.

Percentile Values (χ^2) Chi-Square Distribution with ν Degree of Freedom (shaded area = p)

ν	$x^2_{.995}$	$x^2_{.99}$	$x^2_{.975}$	$x^2_{.95}$	$x^2_{.90}$	$x^2_{.75}$	$x^2_{.50}$	$x^2_{.25}$	$x^2_{.10}$	$x^2_{.05}$	$x^2_{.025}$	$x^2_{.01}$	$x^2_{.005}$
1	7.88	6.63	5.02	3.84	2.71	1.32	.455	.102	.0158	.0039	.0010	.0002	.0000
2	10.6	9.21	7.38	5.99	4.61	2.77	1.39	.575	.211	.103	.0506	.0201	.0100
3	12.8	11.3	9.35	7.81	6.25	4.11	2.37	1.21	.584	.352	.216	.115	.072
4	14.9	13.3	11.1	9.49	7.78	5.39	3.36	1.92	1.06	.711	.484	.297	.207
5	16.7	15.1	12.8	11.1	9.24	6,63	4.35	2.67	1.61	1.15	.831	.554	.412
6	18.5	16.8	14.4	12.6	10.6	7.84	5.35	3.45	2.20	1.64	1.24	.872	.676
7	20.3	18.5	16.0	14.1	12.0	9.04	6.35	4.25	2.83	2.17	1.69	1.24	.989
8	22.0	20.1	17.5	15.5	13.4	10.2	7.34	5.07	3.49	2.73	2.18	1.65	1.34
9	23.6	21.7	19.0	16.9	14.7	11.4	8.34	5.90	4.17	3.33	2.70	2.09	1.73
10	25.2	23.2	20.5	18.3	16.0	12.5	9.34	6.74	4.87	3.94	3.25	2.56	2.16
11	26.8	24.7	21.9	19.7	17.3	13.7	10.3	7.58	5.58	4.57	3.82	3.05	2.60
12	28.3	26.2	23.3	21.0	18.5	14.8	11.3	8.44	6.30	5.23	4.40	3.57	3.07
13	29.8	27.7	24.7	22.4	19.8	16.0	12.3	9.30	7.04	5.89	5.01	4.11	3.57
14	31.3	29.1	26.1	23.7	21.1	17.1	13.3	10.2	7.79	6.57	5.63	4.66	4.07
15	32.8	30.6	27.5	25.0	22.3	18.2	14.3	11.0	8.55	7.26	6.26	5.23	4.60
16	34.3	32.0	28.8	26.3	23.5	19.4	15.3	11.9	9.31	7.96	6.91	5.81	5.14
17	35.7	33.4	30.2	27.6	24.8	20.5	16.3	12.8	10.1	8.67	7.56	6.41	5.70
18	37.2	34.8	31.5	28.9	26.0	21.6	17.3	13.7	10.9	9.39	8.23	7.01	6.26
19	38.6	36.2	32.9	30.1	27.2	22.7	18.3	14.6	11.7	10.1	8.91	7.63	6.84
20	40.0	37.6	34.2	31.4	28.4	23.8	19.3	15.5	12.4	10.9	9.59	8.26	7.43
21	41.4	38.9	35.5	32.7	29.6	24.9	20.3	16.3	13.2	11.6	10.3	8.90	8.03
22	42.8	40.3	36.8	33.9	30.8	26.0	21.3	17.2	14.0	12.3	11.0	9.54	8.64
23	44.2	41.6	38.1	35.2	32.0	27.1	22.3	18.1	14.8	13.1	11.7	10.2	9.26
24	45.6	43.0	39.4	36.4	33.2	28.2	23.3	19.0	15.7	13.8	12.4	10.9	9.89
25	46.9	44.3	40.6	37.7	34.4	29.3	24.3	19.9	16.5	14.6	13.1	11.5	10.5
26	48.3	45.6	41.9	38.9	35.6	30.4	25.3	20.8	17.3	15.4	13.8	12.2	11.2
27	49.6	47.0	43.2	40.1	36.7	31.5	26.3	21.7	18.1	16.2	14.6	12.9	11.8
28	51.0	48.3	44.5	41.3	37.9	32.6	27.3	22.7	18.9	16.9	15.3	13.6	12.5
29	52.3	49.6	45.7	42.6	39.1	33.7	28.3	23.6	19.8	17.7	16.0	14.3	13.1
30	53.7	50.9	47.0	43.8	40.3	34.8	29.3	24.5	20.6	18.5	16.8	15.0	13.8
40	66.8	63.7	59.3	55.8	51.8	46.6	39.3	33.7	29.1	26.5	24.4	22.2	20.7
50	79.5	76.2	71.4	67.5	63.2	56.3	49.3	42.9	37.7	34.8	32.4	29.7	28.0
60	92.0	88.4	83.3	79.1	74.4	67.0	59.3	52.3	46.5	43.2	40.5	37.5	35.5
70	104.2	100.4	95.0	90.5	85.5	77.6	69.3	61.7	55.3	51.7	48.8	45.4	43.3
80	116.3	112.3	106.6	101.9	96.6	88.1	79.3	71.1	64.3	60.4	57.2	53.5	51.2
90	128.3	124.1	118.1	113.1	107.6	98.6	89.3	80.6	73.3	69.1	65.5	61.8	59.2
100	140.2	135.8	129.6	124.3	118.5	109.1	99.3	90.1	82.4	77.9	74.2	70.1	67.3

Source : E.S. and H.O. Hartley, Biometrika Tables for statisticaians, Vol. 2 (1972) Table 5, page 178, by permission called from 'Statistics" 3rd Edition Shaum's Outline Series

BIONOMIAL COEFFICIENTS

N	Err	Err	Erro	Erro	Erro	$\binom{n}{5}$	$\binom{n}{6}$	$\binom{n}{7}$	$\binom{n}{8}$	$\binom{n}{9}$	$\binom{n}{10}$
0	1										
1	1	1									
2	1	2	1								
3	1	3	3	1							
4	1	4	6	4	1						
5	1	5	10	10	5	1					
6	1	6	15	20	15	6	1				
7	1	7	21	35	35	21	7	1			
8	1	8	28	56	70	56	28	8	1		
9	1	9	36	84	126	126	84	36	9	1	
10	1	10	45	120	210	252	210	120	45	10	1
11	1	11	55	165	330	462	462	330	165	55	11
12	1	12	66	220	495	792	924	792	495	220	66
13	1	13	78	286	715	1287	1716	1716	1287	715	286
14	1	14	91	364	1001	2002	3003	3432	3003	2002	1001
15	1	15	105	455	1365	3003	5005	6435	6435	3005	3003
16	1	16	120	560	1820	4368	8008	11440	12870	11440	8008
17	1	17	136	680	2380	6188	12376	19448	24310	24310	19448
18	1	18	153	816	3060	8568	18564	31824	43758	48620	43758
19	1	19	171	969	3876	11628	27132	50368	75582	92378	92378
20	1	20	190	1140	4845	15504	38760	77520	125970	167960	164756

VALUES OF e^{-m} (For Computing Poisson Probabilities)

$(0 < m < 1)$

M	0	1	2	3	4	5	6	7	8	9
0.0	1.0000	.9900	.9802	.9704	.9608	.9512	.9418	.9324	.9231	.9139
0.1	0.9048	.8958	.8860	.8781	.8694	.8607	.8521	.8437	.8353	.8270
0.2	0.8187	.8106	.8025	.7945	.7866	.7788	.7711	.7634	.7558	.7483
0.3	0.7408	.7334	.7261	.7189	.7118	.7074	.6977	.6907	.6839	.6771
0.4	0.6703	.6636	.6570	.6505	.6440	.6376	.6313	.6250	.6188	.6126
0.5	.0.6065	.6005	.5945	.5886	.5827	.5770	.5712	.5655	.5599	.5543
0.6	0.5488	.5434	.5379	.5326	.5278	.5220	.5160	.5117	.5066	.5016
0.7	0.4966	.4916	.4868	.4810	.4771	.4724	.4670	.4630	.4584	.4538
0.8	0.4493	.4449	.4404	.4360	.4317	.4274	.4232	.4190	.4148	.4107
0.9	.0.4066	.4025	.3985	.3946	.3906	.3867	.3829	.3791	.3753	.3716

$(m = 1, 2, 3, \ldots 10)$

M	1	2	3	4	5	6	7	8	9	10
e^{-m}	.36788	.13534	.04979	.01832	.00698	.00279	.00092	.000953	000123	000045

Note : To obtain values of e^{-m} for other values of m, use the laws of exponents.
Example, $e^{-2.00} = (e^{-2.00})(e^{-00.35}) = (.135.34)(.7047) = .095374$

Index

Differentiation, 2

Functionn of Functions, 4

Partial Derivative, 9
Product Rule, 7

Quotient Rule, 8